TRANSITIVE MEDIA [PHILOSOPHY]

Exeter Concepts in Media Thinking

Series editors:
Bernd Herzogenrath, Goethe University of Frankfurt am Main and Tomáš Dvořák, Academy of Performing Arts, Prague

Media Philosophy is a new terrain for exploring questions of media. Emerged yet emancipated from traditional Media Studies, media philosophy as understood in this series focuses on media's specific capacities and potentials. Can media do 'thinking'—can media do philosophy? And if the answer is 'yes' — what are the concepts that emerge from these practices?

Exeter Concepts in Media Thinking provides a venue for thinking anew about the relationship between media and philosophy. Can these disciplines benefit from each other, and—more provocatively, perhaps—do film and its older sibling arts/media (literature, music, fine art, photography, etc.) amount to forms of thought that are partially independent of philosophical reflection 'proper' and thus potential rivals to it? That is to say, are artistic practices and their visual, auditory, and textual expressions *on the same footing* as philosophy, considered as a *thinking* practice? The focus of this series does not consist in finding definitions, obtaining empirical investigations, or assembling a definite critique of the phenomenon 'media'. It rather searches for philosophically relevant reflection, particular *medial* philosophical ways of thought: specifically medial (cinematic, sonic, poetic, etc.) experiments with new forms of philosophical expression of what media 'can do'.

Other titles in the series:

Ontographies: A Media Philosophy of Immanence
Lorenz Engell

Modulation: The Parameter-Image
Tomáš Dvořák

Phenomenotechnical Togetherness
Mark Hansen

TRANSITIVE MEDIA [PHILOSOPHY]

Think(er)ing Technologies in Real-Time

WOLFGANG ERNST

UNIVERSITY
of
EXETER
PRESS

First published in 2026 by
University of Exeter Press
Reed Hall, Streatham Drive
Exeter EX4 4QR, UK

www.exeterpress.co.uk

Transitive Media [Philosophy]: Think(er)ing Technologies in Real-Time

Exeter Concepts in Media Thinking
ISSN 2977-8956 Print
ISSN 2977-8964 Online

A CIP catalogue record for this book is available from the British Library

https://doi.org/10.47788/IKLB9885

ISBN 978-1-80413-232-6 Hardback
ISBN 978-1-80413-233-3 ePub
ISBN 978-1-80413-234-0 PDF

EU Authorised Representative: Easy Access System Europe – Mustamäe tee 50, 10621 Tallinn, Estonia, gpsr.requests@easproject.com

Cover concept by Clifford Hayes, designed by Lyn Davies, based on Figure 1.2: Electromagnetic relay as operative electric circuit.

Typeset in Adobe Caslon Pro by S4Carlisle Publishing Services, Chennai, India

Contents

Illustrations

All photographs (including copyright) come from the Media Archaeological Fundus (MAF) of the Department of Media Science at Humboldt University of Berlin. In accordance with this book's core argument, some of these technical items are preferred to be shown in signal-processing being rather than as inactive artefacts.

Acknowledgements

The author is grateful to the series editors Bernd Herzogenrath and Tomáš Dvořák for inviting him to submit a proposal for their intellectually refreshing *Exeter Concepts in Media Thinking* book series. Further thanks go to Commissioning Editor Becky Taylor and her team at University of Exeter Press for their hospitable and patient accompaniment of the editing process from the very beginning to the final publication. The Faculty of Cultural, Social and Educational Studies at Humboldt University of Berlin generously granted the author a sabbatical in summer 2023 to reorganize the Media Archaeological Fundus in the Department of Media Science, which initially induced the author's practical experience and cognitive impulse to consider a more *transitive* media philosophy. Thanks as well to my former student assistants Raphael Tostlebe and Constantin Matti Roth for photographing the items from the Media Archaeological Fundus that 'illuminate' this text. Further thanks go to the reviewers for sympathetic suggestions both to improve scholarly arguments and in terms of style. It is a privilege that the present book series format allows for a rather manifesto-like mode of argumentation. For the abundance of idiosyncrasies, neologisms and aphorisms in the present text the author takes full responsibility. The 'exile' for a final polishing of this essay has been made possible by the assistance of Ksenija Cerce.

W.E., October 2025

Prelude

Philosophizing in the Presence of Media Technology 'on Stage'

Writing 'about' Transitive Media Philosophy: An Oxymoron?

The marvels of ubiquitous digitalization in communication and information techniques are currently resulting in a breathtaking sophistication of machines. This goes along with a shift of emphasis that increasingly favours the awareness of software (alphanumeric *lógos*) against the still-inevitable hardware (material *technē*). A new reading of the very term techno*logy* itself is necessitated. Direct experience of the machine is replaced by abstract coding on the programming side, while to most media users the technology behind the human–machine interface becomes opaque. The more human users seem literally 'in touch' with communicational devices, the more their very technical essence recedes into an almost non-intelligible sphere. This retreat culminates in the 'hidden layers' of 'deep' neuronal nets in artificial intelligence. A more *transitive* media philosophy will be proposed here—in the grammatical sense of an immediate relation between verbal expression and its objects—where the human actor remains consciously entangled in the technical action.[1] Its aim is keeping actual technologies transparent for critical but most of all techno-philosophical cognition already *while* they are in action. This mode of attention focuses on *instantaneous* technical insights which are at the same time epistemic—that is, deeply knowledge-oriented. A multitude of meanings of the 'transitive' in the context of media philosophy—ranging from a mental tuning in to concrete technical events themselves—will be unfolded in the following modular arguments. Beyond its specific case studies, this book is not only addressed to specialists in engineering and media science but has a fundamental concern for the general understanding of the current media-cultural situation and the need to philosophically trace the essence of technologies while they are in action.

1. In analogy to Barthes, 'To Write', pp. 141–42.

Technological transparency ordinarily refers to obvious and urgent media-critical concerns such as code criticism and data 'bias', the economics of 'repair culture' or environmental issues like the usage and recycling of rare minerals in technical devices that are linked to precarious production infrastructures and politics.[2] But in addition, the present book's effort is rather to preserve, at the same moment of critical analysis, the inherent philosophical beauty, elegance and serendipity of media technologies and to keep their inner-technical knowledge accessible for immediate insight that is—at least momentarily—liberated from more discursive concerns. Such a media philosophy is 'transitive' in the sense of three aspects that will be discussed: a) an instant awareness of the epistemic surplus that is involved in any technological procedure, b) the inherent continuity of inner-technical signal processing (in defiance of 'the digital'), and c) the necessity of keeping technologies transparent for critical inspection. As indicated in the typography of this book's title, this volume refers to transitive media in the technological sense as well as to a transitive media philosophy that instantly links said media with fundamental questions of knowledge and insight.

With its orientation towards a more transhumanistic, even autonomous concept of technological agency, this book is—mostly implicitly, occasionally explicitly—in dialogue with resonant media-theoretical discussions in the fields of new materialism, speculative realism, actor–network theory, even object-oriented ontology. Most of the insights that will be addressed in the following chapters, though, first of all arose from direct contact with the machines rather than from metatextual mediation. The main objective of transitive media philosophy is to share the author's direct attention to the epistemic value of non-discursive technical assemblages themselves. Its primary impulse is inductive, not deductive. The discursive reintegration of such a media-archaeological experience into related scholarly contexts occurs where it seems appropriate.

Conventional media theory, as a predominantly textual and philological practice, is rarely directly confronted with its real reference: the technical event as such. Distant reflection seems like the precondition for theoretical insight. In most philosophies of technology, the actual apparatus is rather discussed as an object than as an agency in co-presence to critical reasoning. This changes when knowledge-oriented insight switches from the abstracted study to an epistemic mode of knowing media that arises from a *transitive* acquaintance and literally experi*mental* communication with the machine in

2. For such a veritably radical media archaeology see Parikka, *A Geology of Media*.

persistent media-epistemic alert. Scientist Hermann von Helmholtz immediately derived sparks of philosophical insight from empirical experiments with laboratory instruments such as his acoustic resonators, as brilliantly expressed in his programmatic monograph *On the Sensations of Tone as a Physiological Basis for the Theory of Music*. Most media artists have a technical medium awareness already during the creative process, and engineers are functionally reflexive of their material practice in real-time. Transitive media enquiry also derives an epistemic surplus from direct entanglement with the medium event itself, both as insight *into* and revelation *from* the technological device. This reciprocal relation is the ontologically unique and media-specific essence of what in this present text is defined as *tec-knowledge*.

This neologism addresses a mode of knowing that does not rise above the machine towards abstraction but remains entangled with its mechanism. An instantaneous and reciprocal dialogue with machines—in the sense of 'cycles of continuous reciprocal causality that connect embodied cognition with the technological infrastructure'[3]—does not only result in a more immediate understanding of technology. To the reflexive mind it additionally reveals the technicity of cognition itself, just as in Henri Bergson's analysis of 'The Cinematographic Mechanism of Thought'.[4] Cinema itself can indeed become epistemically self-reflexive *in* and *by* its own chronophotographic medium (the genre of 'photofilm'[5]). Just like aesthetic—even 'auratic'—experience depends on the radical presence of the artwork, transitive media philosophy is immediately triggered by—and is media-archaeologically rooted in—the medium event itself. This momentum may even be preserved as dynamic short-term memory for later textual explication, like the reverberations or recurrent pulses in a technical delay line.

A particular aesthetic challenge is well known from writing literature and poetry: How closely is the work involved in the world it describes, rather than being just its literally 'symbolic' abstraction? The present text is confronted with a parallel philosophical challenge and apparent oxymoron: Is it possible not only to write media-theoretically *about* technology, but to *write technology* at the same time? Can there be an immediate participative and at the same time reflexive knowledge of technical devices that transcends its unconscious intuitive experience? The temporal mode of such a *transitive* media philosophy is the instant flash of epistemic illumination that co-evolves

3. Hayles, *How We Think*, p. 96.
4. Bergson, *Creative Evolution*, chapter IV, pp. 272–370.
5. Boros, 'Visual and Textual Narratives in the Photofilms of Gusztáv Hámos and Katja Pratschke', p. 221.

from the observer being coupled to the specificity of the technical event itself. Operational 'errors' in particular become media-analytic illuminations here, such as the occasional echo effects in video conferencing: They remind of the actual latencies that arise from the digital communication channel, revealing the computational essence of stepwise signal processing against the phenomenology of continuous 'live' transmission and its logocentric illusion.

Far beyond a simple hands-on approach, transitive media analysis is not regressive but a form of haptic and at the same time reasoned understanding (in the double sense of German *begreifen*), driven by practical technical experience but framed by epistemic curiosity. For this kind of analysis of technological artefacts, media archaeologist Erkki Huhtamo coined the term *thinkering*.[6] While artistic 'thinking in the act' such as dance, or painting,[7] or perceptual 'sounding and listening as a knowing-in-action',[8] is still somewhat body-centric, transitive media epistemology recursively deals with objectifications of the mind that are technically implemented (rather than 'embodied'). Technologically informed matter, down to the artificial neural nets in contemporary machine learning—if not structurally, but materially—differs from the bio*lógos* of evolutionary humankind. Very much in accordance with Katherine Hayles's sense of machine cognition (but perhaps more concrete in terms of engineering[9]), technological devices are not only perceivable but knowable, as a recursive encounter and *self*-reflexive feedback of a philosophical mind that is glued to the signals of technically informed matter—just like semiotician Roland Barthes felt 'glued' to the moving image.[10] In mechanisms (rather than organisms) a both inward and external(ized) thought is already techno-logically *given* and can, in return, be derived from within the agency of the machine.

Reading this Text: The Transitive Media Momentum Is Now

In order to transgress the inherent contradiction of *intransitively* writing about a claimed *transitive* media philosophy by way of a distancing theory, the argumentation has to stay aware, at every moment, of its technical status as symbol processing. Its aim is to *write the medium* reflexively. At first glance,

6. Huhtamo, 'Thinkering with Media', pp. 33–46.
7. Manning and Massumi, *Thought in the Act*, p. vii.
8. Feld, 'Acoustemology', p. 12.
9. See chapter 5, 'Cognitive Assemblages: Technical Agency and Human Interactions', in Hayles, *Unthought*, pp. 115–41.
10. Barthes, 'Leaving the Movie Theater', p. 348.

the present text seems to negate this claim, since its printed strings of characters are no longer in touch with the electronic means that have enabled their production. But its writing process has been aware, at every moment, of the difference between typographic printing and electronic writing as media operations themselves. Whereas the material nature of all the words in a print text is immediately transparent to the media-conscious mind, electronic texts intransitively hide their internal mode of existence. 'Code always has some layers that remain invisible and inaccessible to most users.'[11] The transient media philosopher, though, as author is steadily aware of the 'symbolical machine' in its literal sense. In ancient Greece, the notation of philosophy or poetry, as well as musical scores and mathematical notation, actually occurred in one and the same alphabet of discrete symbols (*stoicheia*).[12] Such coding did not simply serve functional recording purposes but was simultaneously experienced as a techno-logical form of *operative world analysis*. Actually *doing thought*, nowadays known as 'enactivism' in cognitive science,[13] enhances the mechanistic (rather than organicistic) model of reasoning and intelligence as symbol manipulation.

Most of the technological scenes that are referred to in this book necessarily stay outside of this text. Alphabetic characters can describe and represent but never actually emulate a material device like the electromagnetic relay. But the very writing scene itself comes close to transitive media philosophy once the reader does not only focus on the semantic referential content. By reading this text *literally*, the reader becomes simultaneously aware of the medium message of typographic expressions on material paper (or electronic interfaces). The average usage and consumption of technologies can indeed be entangled with their conscious experience as autonomous agencies. As objectifications of the mind, they oscillate between media function and medium knowledge. Most vanguard poets and literary writers have been conscious of the materiality of their writing tools. The actual medium message of the alphabet, notwithstanding its semantic content, is the mechanism of a symbolic *ars combinatoria* that has been reflected from ancient (and subsequently baroque) times down to the Transformer deep learning architecture underlying dialogic artificial intelligence like ChatGPT today, with its predictive stochastics. Differing from the hermeneutic focus of writers and readers on thematic content, transitive media attention additionally is aware of the mechanisms of the symbolical code (alphabetic software) on the one

11. Hayles, 'Print Is Flat, Code Is Deep', p. 75.
12. Lohmann, *Musiké*, pp. 10–11.
13. Notably Clark and Chalmers, 'The Extended Mind', pp. 7–19.

hand, and the materiality of ink on paper (hardware) on the other. As long as textual expressions become processed in the human mind exclusively, this cultural technique remains a monologue, different from human interaction with operative technologies such as computer programming. The word does not yet become flesh (to quote John 1:14) with only ink on paper as a passive storage medium. The abstract digital bit ('0/1'), in its information engineering sense, on the other hand, is self-active once it becomes an articulated 'word' (*lógos*) in techno-logically informed and energetically biased matter, such as the electromagnetic relay.

The technologically conscious reader is reminded that, *at this very moment of reading*, they (and the unconscious decoding mechanism called 'it') are subject to a decoding of printed letters (or luminescent pixels) that turn any understanding into the function of a symbol-processing machine. Turing himself had such a moment. In his seminal paper 'On Computable Numbers' which has since been quoted, in media historiography, as the conceptual foundation of the digital computer, his argument is interrupted by the insight that any apparently ideal algorithm may be vetoed by the very material limitations and idiosyncrasies of pencil and paper as its writing scene.[14] Being necessarily embodied (in humans) or entangled with supporting alternative matter (in electronics) in order to unfold at all, *lógos* is always already *im*pure—be it computation as mental (that is, electro-chemical) thought experiment, or abstract notation on paper. Once observed closely on the electronic-processing level, even digital *computation* is revealed as analog *computing*. The present participle accentuates that a technical device is in its media-being only as a signal-transducing and -processing entanglement of reason with matter.

Binary computing, as mathematics and logic made operative, is never numerically autonomous but transitively glued to physical substance. According to Turing's definition of the algorithm, the symbolic *is* the machine already. Turing conceived the algorithm not simply as an instrumental pre-condition for actual computing. Alphanumerical reasoning is revealed as *mechanism*, as also defined by Peirce: 'every machine is a reasoning machine, insomuch as there are certain relations between its parts, which relations involve other relations that were not expressly intended. A piece of apparatus for performing a physical or chemical experiment is also a reasoning machine,

14. 'If we regard a symbol as literally printed on a square [...]': Turing, 'On Computable Numbers', p. 249 (footnote). This also recalls Roman Opałka's material paintings of numbers from one to infinity on canvas, starting from 1965 at his studio in Warsaw, titled *1965 / 1 – ∞*. Thanks to Ksenija Cerce for this notification.

with this difference, that it does not depend on the laws of the human mind, but on the objective reason embodied in the laws of nature.'[15] Technological reason is not simply symbolically imposed upon the real. There are self-expressions of *lógos* from matter itself. Such a momentary insight may be elaborated media-philosophically and trigger metaphysical thoughts (in the Aristotelian sense of 'first principles') about the entanglement of 'pure reason' with sheer matter and energy.

How to 'Illustrate' the Technical Momentum Transitively

A similar flip from an abstract (intransitive) to a transitive (concrete) media experience is performed in relation to the illustrations in this book. The images of technical devices, as frozen representations, cannot reveal the processual dynamics from which media-epistemic illuminations arise. But the reader may become aware of the reproduction scene itself. While chemical photographs still carry an indexical trace of the physical optical contact with the technical device they present, digital reproduction (pixels) turn images into symbolic abstractions of their information value. Once the reader does not only see the iconic reference but becomes aware of the *imaging* scene itself, a transitive medium experience unfolds. The very term *illumination*, in media-archaeological materialism, stems from medieval manuscripts that frequently incorporated gold leaf on the page. In analogy to religious mysticism in medieval and baroque times, transitive media insight occasionally arises from such 'illuminative' moments. Once tuned to media-epistemic attention, the reader might not only 'see' the circuit diagram or algebraic expression illuminate, for instance, an electromagnetic trigger relay. The reader may also become aware of its mate*realization* (to allow for a typographic idiosyncrasy) as objectification of fundamental philosophical concerns like the entanglement of 'pure' reason *with* (or even *as*) the physical impure world. Such issues have been discussed in occidental culture since Plato's philosophical idealism and in theological concerns with an abstract God that becomes incarnated on earth. With material ink on paper in typographic printing (or alternatively with the electric voltage of pixels on the material computer screen), the entanglement of *lógos* and matter itself actually takes place.

Once the complicated layers of techno-mathematical knowledge have been disclosed and declassified from the technological archive and its concrete material or energetic instantiations have been identified, they can be epistemically reflected and extended to media-ontological insights. In transitive

15. Peirce, 'Logical Machines', p. 630.

media philosophy this happens not only sequentially from empirical (practical) experience to theoretical (textual) distance but as primordial interference. All of a sudden, even printed electronic engineering diagrams are no longer passive illustrations but become transitively enacted by the philosophically cooperative mind. For example, a binary switch (found for instance in early electromechanical computing) is not only deciphered in arithmetic or logical terms ('0/1', 'true/false') but can also be epistemically considered in its concrete temporeality,[16] pondering the micro-temporal delay that occurs between two states. In a significantly different media-epistemic mode this even applies to the synchronous 'non-time' qubit sequences in quantum computing. Discrete data processing is revealed as truly analog once 'instant' logic comes into being in an always impure and thermodynamically time-consuming material world.

Theoría vs 'Thinkering': Introducing the Media Archaeological Fundus

The style of the present argumentation may occasionally look declarative in its thetic insistence. But its staccato dramatizations are actually grounded in a sound experience of the *transitive* media-philosophical concern: epistemic investigations of the concrete techno-logical encounters between knowledge and matter. Media theory usually gains its epistemic strength from a distance, as a symbolic (textual) post-production of real (electro-)technical events by means of recombinant reflections. Instant tec-knowledge, on the contrary, arises from a direct coupling of human cognition and the medium event.[17] The technical object is experienced as co-subjective agency. At the same time the fundamental dichotomy between human and non-human agencies is both acknowledged and permanently questioned. Different from the biological model, the premise of a truly cybernetic human–machine symbio*techno*sis actually gains a knowledge surplus and ontological insights from the frictions that occur in the coupling of the human (biologically embodied) and technically implemented *lógos*.[18] This situation is analogous to the claim that

16. This deliberate typographical neographism expresses a technically concrete (rather than abstract transcendental) understanding of 'time'.

17. In this regard, the 'cold' media-archaeological gaze on technological devices is premised on a (at least momentary) distance to discursive concerns such as their embedding in social and other contexts and environments, in favour of a most passionate techno-epistemic immersion.

18. See Licklider, 'Man–Computer Symbiosis', pp. 4–10.

'there can be no proper theory of techno [music], just a practical philosophy of electrified bodies'.[19]

When the turntable of an obsolete record player, with a retired motor, is switched on after a long time of non-usage, it may be that it hardly moves at all. Only after a certain delay will it slowly start to accelerate. For human ears a singing voice that has been engraved on the record thereby emerges from initial silence to a subsequent unrecognizable noise and finally to a phonic signal: becoming *lógos*. This may even become tragicomical if the vinyl LP recording of philosopher of technology Martin Heidegger's lecture *Die Kunst und der Raum* is the record played.[20] The phonographic reproduction medium turns into a sound-analytical device once its literal *understanding* switches from mere aesthetic consumption to an explicitly 'acoustemic' enquiry in a media-ontological radicalization of this originally anthropological term.[21] The whole range of the sonic signal-in-being that unfolds from the scratch in the vinyl groove to music as recorded content is revealed by the apparently erroneous medium event as a correlation between the ephemerality of sound and its material embodiment. With noise-cancelling tools in current machine learning, though, the symbolical code transcends even such articulations of the mate*real*, silencing them in favour of medium intransitivity.

While media theory is predominantly a logocentric textual symbolization (a literal 'ABCtraction') or even tokenization from the technical event, transitive media philosophy navigates *within* the signal field. Analogous to this, Feld's acoustemology has transformed 'an anthropology of sound into an anthropology in sound', thinking 'more through recording and playback' than by symbolic representations of knowledge.[22] This media turn from theoretical coding to immediate reflexive knowledge is not simply a critique of the representation model in cognitive science, nor did it 'arrive conceptually as a result of pure theory or from direct abstraction'.[23] It has rather been induced by Feld's ethnographic field studies of songs and sound in the Bosavi rainforest region of Papua New Guinea. By analogy (but with more attention to the difference between a phenomenology of perception and explicit media-archaeological analysis) the moments of transitive media philosophy

19. 'Es kann in diesem Sinne keine eigentliche Theorie des Techno geben, sondern nur eine praktische Philosophie der electrifizierten Körper.' QRT, 'Tekknologic als Tekknowledge', p. 28 (translation author's own).
20. Record number 30–962 (St Gallen: Erker Verlag, 1989).
21. This neologism has been coined by Steven Feld. See Feld, 'Acoustemology', pp. 12–21.
22. Feld, 'Acoustemology', p. 18.
23. Feld, 'Acoustemology', p. 15.

in this present study are no abstract or simply conceptual claim but arose from a concrete media-theatrical scene: an extended experience inside the Media Archaeological Fundus (MAF) at the Department of Media Science at Humboldt University of Berlin during a practice-based research sabbatical in summer 2023.[24] The MAF (where the author acts as Senior Curator) is actually located on the site of philosopher G.W.F. Hegel's former house. Media-philosophical musing in the presence of the MAF's technical devices is literally 'inspired' (as well as challenged) by its house ghost but with the notable difference that essential thoughts and scientific insight, in the case of technologies, are not 'only to be won through the labour of the Notion' but through media in operative being.[25]

For the present book, all illustrations have been chosen from this epistemically oriented assemblage (rather than merely media-historical collection) of various mechanical, electromechanical and electronic-processual artefacts (both 'analog' and 'digital') across techno-logical times. An assemblage, in Gilles Deleuze's and Félix Guattari's sense (French *agencement*), is 'a rhizomatic and fragmentary multiplicity that never adds up to "One"'. Differing from their concept of the abstract machine, an assemblage comes close to the epistemic concreteness of transitive media philosophy: 'While linguistically oriented translations rendered it as "concatenation of enunciations", it gradually became clear that it was actually about the complex linkages between sign and matter, technology and body'.[26]

Beyond bare media museology, the MAF's aim is to provide an experience that inspires transitive thinking and tinkering *with* (not simply *about*) technological core elements regarding their interrelated knowledge implications. In order to compose not only a technically but also epistemically focused inventory ('Kata*lógos*') of artefacts, the author found himself, for an extended time span, literally *within* media hardware instead of their external textual research and verbal expression. The thematic rearrangement of technical artefacts was accompanied by practically testing them. According to the 'Berlin school' of media science's premise, technical objects are in a media condition only when *in being*—that is, operating and processing signals. The inevitable challenge of hands-on repair, from the perspective of a (non-engineer) media theorist, resulted in flashes of micro-epistemic insight and literal short circuits between the epistemically curious amateur technician

24. https://medienwissenschaft-berlin.org/medienarchaeologischer-fundus
25. Hegel, *Phenomenology of Spirit*, p. 43 ('Preface').
26. Herzogenrath, '"Wait a Minute"', p. 1032, https://dx.doi.org/10.1353/mln.2022.0074

and the technical object. In such moments of epistemic transitivity, the observational distance implodes, resulting in a 'tunnelling' (in its quantum-physical and communication-engineering sense) of insight between the operating mind and the media agency. What has been described as 'enactivism', beyond the 'embodiment'-focused anthropocentrism in classical cognitive science, becomes reciprocal in such human–machine couplings.

For the close participant observer of technological action, a weak subjectivity and non-emphatic idea of the human individual is required in order to immediately adapt to the media-epistemic articulation that occurs during the time slot that opens in an intense coupling between mind and technology. In such moments of mutual forms of reasoning, an almost 'affective' awareness of the media-philosophical momentum arises. While the affect is usually defined as a preconscious neural moment, for transitive media philosophy the epistemically conscious affect is no oxymoron. It rather names a state when the medium-specific process has not yet receded into the past and is only then reflected from belated temporal and conceptual distance as textualized 'theory'. Unlike a personal diary or a technical logbook, transitive media philosophy is aware of the media-epistemic momentum on the fly. While for Foucault the 'archive' (*l'archive*) of an actual practice becomes describable only in the moment when this operative present recedes into the past, transitive media epistemography is already involved in the technical event while it happens. Transitive media philosophy thereby becomes confronted with the challenge of relentless adaptation to new technical constellations with the accelerating speed of techno-logical change itself. Differing from the theatrical elevation that usually keeps the observer distant from the actual stage or orchestra of events (reminding of the etymological link between *theatron* and *theoría*), transitive media philosophy is a form of reflection that arises from immediate human–machine interaction.

Delayed transitive media philosophy carries over that immediate impulse into a subsequent writing act to maintain such insight. But it does not completely subject the more distanced theoretical text to the symbolic order, but couples it back to the technical real by further feedback—that is, by continued experimentation with the apparatus in action. Transitive media philosophy thereby insists upon the grounding of media analysis in the actual technological event. Postponed textual reflection is usually cut off from the essential definition of media. In contrast to other rather immobile cultural artefacts, a technical device is in its medium state only when transducing signals. The transitive approach therefore does not refer to the material technical artefact as distant theoretical object, but to the immediate experience of its processuality. The test *of* and *for* technical media reasoning is its confrontation with the actual machine in its effective operation. Notably, a transitive epistemology of the computer, in its techno-mathematical double-bind, does not reduce this mechanism to a pure thought experiment (Turing's symbolic *computation*) but encompasses its actual *computing* as well (electronic or otherwise). While in the mathematical model of algorithmic procedures computational reasoning has been logocentristically defined as an abstract 'Turing machine', a different kind of machinery emerges where the relation between *lógos* and the matter it acts upon is no longer intransitive but interacts with its materiality. The concept of 'deep' machine learning is still modelled by algebraic formulae and numerically abstract data. But what if *lógos* finally unfolds from the operative technified *matter* of artificial neural nets themselves?

Getting in Touch with the Actual Artefact

The technical artefacts are preserved in latency on the shelves of the MAF for enactment as knowledge agencies. The central workbench actively invites a media anatomy and introspection that does not primarily serve for repair but for technical experience that is instantly reflected through an epistemologically biased cognitive filter. In contrast to the textual reasoning of media-theoretical expression, transitive media philosophy listens to the peculiar articulations of the technical device with epistemically tuned mental ears. Technical articulation requires the actual material medium to come into being, just like a tone unfolds as an extended presence that can only be quoted symbolically. Media-scientific knowledge 'priming' (to borrow a term from neurobiology) allows for the philosophically oriented mind to structure a recognized technical phenomenon into an epistemic insight: pensively but while it is occurring. Through cognitive human–machine coupling this happens in real-time (understood here in its engineering sense of a well-defined temporal frame of presence), just like a musical pattern can be derived from listening to an otherwise merely acoustic sequence.[27]

Occasionally, failures and defects of artefacts from the MAF occur in the moment of their activation for *demon*stration. This is the moment when the item's inner-technical 'demon' is revealed and set free, which can be media-epistemically recognized as a form of techno*lógos* that even extends to apparent *a*logic noise. This momentum from the laboratory work bench is then immediately registered in a media-archaeographical notepad. Like the aesthetics of 'patching' sonic configurations with an analog acoustic synthesizer,[28] the MAF experience insists on grounding deductive

27. Spitzer, *Musik im Kopf*, p. 131.
28. As demonstrated in Bjørn, *Patch & Tweak with Moog*.

media-theoretical insight in immediate contact with the electronic instrument. A concrete transitive media phenomenology results from reactivating dormant past devices, such as a magnetophonic steel tape recorder: 'rather than a purely epistemological reverse engineering, [...] a concrete redoing of the material artifact [...] allowed us to get closer to steel's very own audible past'.[29] This reties transhistorical electronic circuitry with its material archaeology and induces a departure from distant intransitive writing *about* technical circuits towards direct media research *with* circuits themselves.[30] Technical re-enactments of the past thereby become *operative* media dramaturgy that takes place with(in) the technological device itself.

In an extended sense of McLuhan's notorious theorem, not only a hidden technical 'medium message' but also the option of conscious philosophical insights is latent behind an exclusive focus on media content (figure against ground, in terms of *Gestalt* theory). Transitive media attention reveals such subliminal tec-knowledge explicitly. Media science thereby differs from communication studies in its focus on the techno-logical impact in itself that alters the cultural subconscious: 'For the "message" of a medium or technology is the change of scale or pace or pattern that it introduces into human affairs.'[31] In traditional concert halls, certain members of the audience (apart from the conductor and musicians themselves) actually read the score while listening to the musical event. In a radical concretization of this cultural technique, Kittler's vision of ultimate media aesthetics is the ability to immediately recognize the electroacoustic synthesizer circuitry itself when listening to electronic music, identifying the technical diagram that is shining (and sounding) through: 'Whosoever is able to hear or see the circuits in the synthesized sound of CDs or in the laser storms of a disco finds happiness.'[32] Transitive media philosophy transforms this aesthetics from literal *machine listening* into an acoustemic insight, short-circuiting the electrical wiring and/or coding with the epistemic reflections of sonic media as such.

The usually invisible generic algorithm becomes aware and transparent as projection on a screen in algorithmicized dance events. 'Algoraves' are live coding competitions where the figure (as optical or acoustic interface) does not hide the medium ground. Computation's proper techno*lógos* thereby becomes media-archaeologically transparent, bringing the technical operations from off-stage to on-stage in real-time. In the Algorave format, the

29. Tkaczyk and Dörfling, 'Wireless Wiring' (forthcoming).
30. 'Forschen *mit* Schaltungen': Dörfling, *Der Schwingkreis*, p. xvi.
31. McLuhan, *Understanding Media*, p. 8.
32. Kittler, *Gramophone, Film, Typewriter*, p. xli.

audience actually watches 'musicians programming live on stage as if they were playing musical instruments.'[33] In such a 'live coding' practice, the processual real-time convergence of abstract algorithm (computation) and its actual rhythmic machine implementation (computing) becomes media-epistemically evident. A fervent machine-oriented ontology insists that '*[e]very* machine has its internal form of temporality and these temporal rhythms differ among themselves'.[34] Media *algorhythmics* also becomes audible via electromagnet wave detectors ('sniffers') that sonify the otherwise implicit sonicity of wireless telecommunication devices that are digitally pulsed.[35] The acoustemically tu(r)ned listener is thereby invited to reconsider the ancient Pythagorean coincidence of musical harmony and mathematics in techno-logical terms. Among the many modes of neural sensation, auditive cognition is most transitive to the awareness of time-critical media execution.[36] While phenomenally providing a kind of immediate 'live' communication with the computing instrument, 'live' coding, though, in fact remains subliminally subject to the time-consuming step-by-step operations of its algorithm. This veritable media theatre radically reminds the human observer/listener of the non-human artefact processuality. Media-epistemic *aisthesis*, instead of intransitive media-phenomenal aesthetics, immediately recognizes the otherness of such a technical chrono*lógos*.

Scientific experimentation (at least in the epoch preceding its computational simulation) has always been aware of the primacy of the material arrangements of the laboratory. The cybernetic notion of a 'generative aesthetics' arose,[37] for instance, from the systemic coupling and feedback between humans and machines. In a similar but more knowledge-oriented mode, the transitive media philosopher experiences epistemic sparks when attending to a technical medium as a conscious experi*mental* [sic] operation. Unlike most familiar philosophy of technology, such a media-epistemic attitude does not abstract from the concrete device but, on the contrary, derives philosophical insight from the most intimate encounter with the real machine or algorithmic mechanism. In analogy to the composer whose knowledge of musical theory

---

33. Miranda and Weaver, 'Introduction to Quantum Computing for Musicians', p. 18.
34. Bryant, *Onto-Cartography*, p. 157.
35. On this neographism see Miyazaki, *Algorhythmics*. For media-epistemic insight that results from the peculiar convergence of beat per minute and rhythmic flow in electronic dance music ('techno'), see Leiner (alias QRT), 'Tekknologic als Tekknowledge', pp. 7–35.
36. On technological time-criticality, see Volmar and Stine (eds), *Media Infrastructures and the Politics of Digital Time*.
37. As most influentially coined by Max Bense, e.g. in 'The Projects of Generative Aesthetics', pp. 57–60.

does not simply precede but instantly interacts with the actual score notation, the transitive media philosopher unleashes fundamental ontological issues such as the infinite variances of the mind–matter dichotomy from the technological devices that are experienced as their processual instantiations. Such flashes occur in real-time and need to be instantly epistemically reflected, since such insight is ephemeral like sound. Transitive media philosophy is radically tempor(e)al. Technological articulations cannot be reduced or abstracted to a distant hermeneutics of the symbolic machine diagram, but require the presence of the wiring diagram to become operationally enacted by its instantiation *as* material and energetic and idiosyncratic machine. While most 'silent' knowledge is embodied in actions and objects and 'not interested in cognition so much as implementation',[38] transitive media philosophy makes so-called implicit knowledge instantly explicit. That makes all the difference to the assertion 'to know something is not to have to think about it'.[39] While the media-epistemically tuned mind turns implicit tec-knowledge into explicit insight by expressing it in philosophical real-time, its proto- (rather than un- or pre-)conscious 'affective' knowledge appeal may even be preserved in latency until it transforms into verbal or textual media theory.

In such experimentations, even a 'historical' or 'dead' media device loses its pastness and temporal distance in favour of an actual, intimate and temporally immediate experience of its technical idiosyncrasies. While in the case of a magnetophone the principle of electromagnetic sound recording is almost timeless, its mechanical materialism grounds the procedure in worldly decay, creating a constant background noise and other frictions such as repaired breaks in the tape. Transitive media philosophy does not simply register but acoustemically listens to mate*real* articulations such as 'uneven magnetization and the resulting fluctuations in signal quality' that do not become evident from reading the patent documents themselves.[40] Magnetic recording is indeed tightly intertwined with its material infrastructure. The (in-)homogeneity of its atomic structure is included in the techno*lógos* hypothesis itself: not only for the sake of emphasizing its environmental impact (as in critical media-ecological discourse) but for pondering related media-epistemic questions such as analog transduction vs digital coding.

Distant media-theoretical visibility differs from more immediate, media-epistemic awareness. While traditional tools are functionally 'ready-to-hand' for human usage, they become 'present-at-hand' to consciousness

38. Giessmann, *The Connectivity of Things*, p. 13.
39. Rancière *The Names of History*, p. 2.
40. Dörfling and Tkaczyk, 'Wireless Wiring' (forthcoming).

only in moments of breakdown.[41] Transitive media philosophy, in analogy to McCulloch's 'experimental epistemology',[42] is aware of the medium's presence even in its regular experience. Once the media theorist becomes practically entangled with a technical instantiation of the logical mind, the enquiring subject is involved in material realities but stays conscious of its epistemic dimension at the same time. Transitive media philosophy recursively occurs in mutual (self-)reflexion between physical *techné* and mental *lógos*. It is such reasonable think(er)ing that differentiates transitive media philosophy from the rather implicit kind of observer–object entanglement in its intra-active correlation, as has been epistemologically generalized by Karen Barad following the Copenhagen interpretation of quantum physics.[43]

Unlike traditional cultural artefacts that are frozen in museum displays, the media-cultural past can be experienced by *re*presencing technical mechanisms as 'temporal objects' themselves.[44] In Humboldt University of Berlin's MAF and Signal Laboratory,[45] apparently 'dead' media devices are operationally re-enacted even at the cost of functionally replicating historically original mechanics and electronics. Their dynamics and agency as media-in-being is prioritized against their silencing through rigid protection of their material authenticity.

In contrast to the rather intellectual objects of philosophical reasoning, the media-cultural past can be experienced transitively by literally getting 'in touch' with the artefacts.[46] In this sense, their inner-technical reasoning is taken seriously. At that moment, the technical device is experienced as a non-human but still knowledge-informed agency, resulting in a post-Platonic dia*techno*logue with the machine that evokes its technical *lógos*. Media-archaeological laboratories nowadays invite us to tinker with technical, especially electronic, devices. Such an experience of medium materiality is vital in times of phenomenally 'immaterial' information flows and 'virtual' reality. If in computer science the 'virtual' denominates entities that exist by algorithmic calculation only (computation), the computing 'lab' experience reminds us of its irreducible roots in electronics. Concerning such techno-mathematics, media-artistic researcher Joost Rekveld actually distinguishes

41. Heidegger, *Sein und Zeit*, p. 69.
42. McCulloch, 'A Historical Introduction to the Postulational Foundations of Experimental Epistemology', pp. 39–50.
43. Barad, *Meeting the Universe Halfway*, p. 139.
44. Husserl, *On the Phenomenology of the Consciousness of Internal Time*, p. 24.
45. On the philosophy of both the Media Archaeological Fundus and Signal Laboratory from a comparative perspective, see Wershler, Emerson and Parikka, *The Lab Book*.
46. Operational videos are available at http://www.youtube.com/user/HUMediaStudies [accessed 14 November 2025].

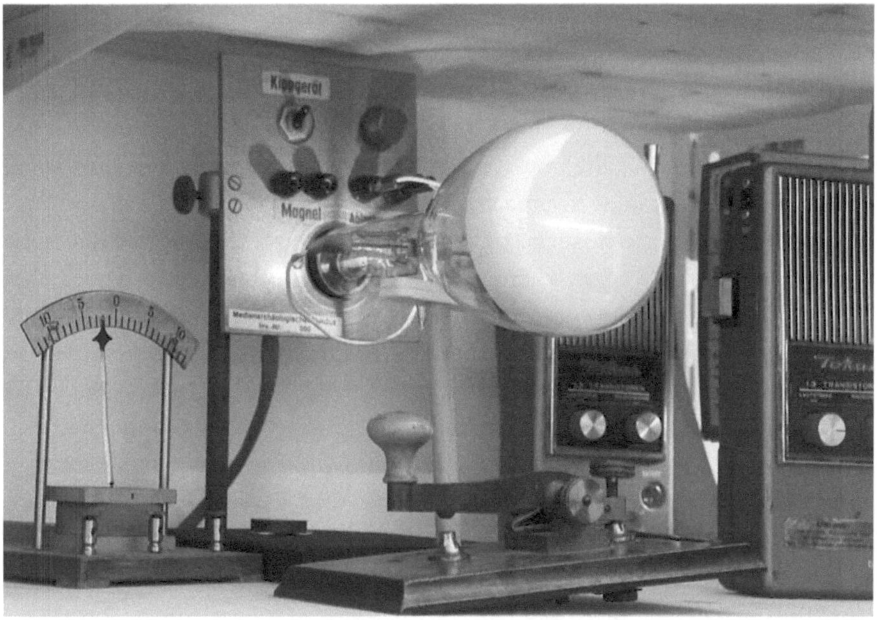

Figure 0.1: The electronic oscilloscope ('Braun's tube') in being: Magnetic deflection of a cathode ray.

between a transitive 'dialogue with machines' and a more intransitive 'dialogue with the mathematics inside of the machines'.[47]

Transitive media philosophy does not only experimentally make implicit techno-knowledge explicit,[48] but grants the technical artefact the role of a co-agency in media thinking itself—just as occurred in the *9 Evenings* media theatre performances in New York in 1966, where the circuit diagrams of electronic devices became the score of the event itself.[49] Material experimentation thereby extends to a revealing of its inherent techno-logics. Such a media-*active* philosophy (instead of philosophy *about* media) does not focus on abstract reason but on its technical groundings or emanations. Computer art may be generated even without explicit aesthetic theory.[50] When Gilbert Simondon philosophically investigated the 'mode of existence of technical

47. Rekveld, *Dialogues with Machines*, p. 49.
48. As it is brilliantly demonstrated in Van den Oever and Fickers,, 'Doing Experimental Media Archaeology: Epistemological and Methodological Reflections on Experiments with Historical Objects of Media Technologies', pp. 45–68.
49. See Clarisse Bardiot, 'The Diagrams of 9 Evenings', https://hal.science/hal-02338052/document [accessed 30 January 2023].
50. 'Um Computerkunst hervorzubringen, ist eine Theorie nicht unbedingt erforderlich.' Franke, *Computergraphik—Computerkunst*, p. 160.

objects', this was no intransitive ontological metareflection. His enquiry into the inherent essence and transductive evolution of technicity is rather inner-technically derived from the precise analysis of, for instance, the electronic specifications of the thermionic vacuum tube. This hardware is not simply understood as the mere basis that shapes superior techno-logical thinking (the materialist approach), but as a form of technical reason that exclusively takes place in such concretizations. The thermionic vacuum tube provides an example of an elementary functional technological device that can be experienced in its epistemic surplus value *by* and *as* transient epistemic t(h)inkering. In electronics, the otherwise natural physical phenomenon of electric current has been *mindfully* domesticated. The vacuum tube, in its triode version, marked the qualitative transformation from simply energetic to rather intelligent (cybernetic) applications of electricity: a logification of energy from within. In media terminology, that makes all the difference between *electric* and *electronic*.

Iterative Efforts to Address the Janus Face of Transitive Media [Philosophy]

Transitive media [philosophy], in its double sense as *transitive media philosophy* and *philosophy of transitive media*, concerns two levels of analysis from which the partition of the chapters in this volume arises: an instant epistemic awareness of technological events, and the close inner-technical analysis of the transitional nature of media themselves, be it continuous aka 'analog' or discretely disguised as the 'digital'. A series of concrete case studies grounds such claims in actual experience, followed by afterthoughts that address the limits of transitive media insight. To what degree is the possibility of a critical analysis, in times of increasingly opaque black-box technologies such as 'deep' machine learning, linked to techno-mathematical transparency? Such technologies may be granted the capacity to themselves establish a transitive relation to media philosophy.

The core issues of such a bidirectional transitive media philosophy are addressed by the present text in a style that is far from a linear (causal and logical) unfolding of arguments. This allows for a discontinuous reading where each subtitled module contains a glimpse of the overall attempt. This book's recursive, sometimes even repetitive pattern and rhythm results from iterative efforts to circumscribe, by various verbal configurations, the oxymoron of a knowledge-oriented immersion in technology that is simultaneously epistemically reflexive and intuitive.

Chapter 1

From Beyond: Transitive Media Insight

Towards a Process-Oriented, Transitive Ontology of Technical Media

This treatise is methodologically consequential to the techno*lógos* hypothesis.[1] A non-anthropocentric philosophy of technology recognizes its inherent epistemic relevance from within and, instead of asking 'What is a medium?', it explores its operative ontology. After all, the *on* in ancient Greek *ontología* is already a temporalizing linguistic unit: not indicating 'to be', but a processual 'in being'. The transitive coupling of machine operativity and performative human 'thinkering' leads to an inseparability of what media *are* and what they *do*. The short-circuiting of ontological questions and pragmatic experience results in a veritable praxeology that does not simply concern media applications but media reasoning itself. Metaphysics becomes physical in/as technology.

Seeing through a physical machine or symbolic mechanism for transparency is still distant, with the optical perspective dominating this concept. An alternative mode of media(-active) contemplation is absorptive reflection, allowing for an immediate knowledge-oriented dia*techno*logue with the artefact. Such a short-circuiting between the medium event and the epistemic mind techno-philosophically results in transitive intelligence: a momentary suspense from the conventional dichotomy between the vibrational sense-affect and cognitive reason—even if such a deconstruction, when formulated in a media-theoretical text, still invigorates this Cartesian dichotomy by repeating its terms.

Such an approach to media considers itself still as philosophy, since it concerns the epistemic, knowledge-focused attention. But its understanding of the *lógos* in technology is not extended to an 'absolute spirit' but rather a

1. On this heuristic theorem see Ernst, *Technológos in Being*.

different emphasis of techniques as 'objective mind'.[2] *Media-specific* philosophy is resonating with techno*lógos* in its sense as operative and *objectified* reasoning—not only as an artificial extension of the human mind but of an *alien*, essentially techno-logical think(er)ing.[3] While mental reasoning is a neural representation, its technical simulations make neural action immediately experimentable. 'One can easily implement our model into an analog circuit, and utilize it as one of the components of realistic artificial neural networks.'[4] Any technical reification (the taking of a hypothetical concept for real) is not metahistorically abstract like the ancient concept of *lógos*,[5] but it is always already 'situated': first of all in concrete materializations that precede discursive contextualizations like the social or the political. A transitive philosophy of technology focuses on the latent epistemic agency of the machine itself—that is, the kinds of insight than only occur techno-logically. With such a quest for an instant media epistemology, technical operations are not simply perceived functionally but as structures that are both informed *by* technical reason and in return inversely trigger techno*lógos* itself. This kind of media-specific experience cannot arise from a distance (*theoría*) but only from the immediate entanglement of the enquiring mind with the machine.

Let There Be Medium Immediacy: Transitive Technological Insight from Real-Time Experimentation

The present approach to technological analysis is not another media theory. It rather oscillates between an objective understanding of transitive media processes and a subjective transitive philosophizing *with* such technological events. Philosophy is, in this context, understood in its scientific sense as decisively *epistemic*—that is, knowledge-oriented reasoning and practice. 'Transitive media theory' would be a contradictory expression, since theorizing is a distancing act like an observation in the optical perspective. Transitive technological insight, in terms of a processual time-critical theory,[6] and with its oxymoronic formula of thinking *media* im*media*tely, rather relates to its

2. These are obvious modifications of key notions in Hegel, *Phenomenology of Spirit*.
3. See Gotthard Günther, 'Metaphysik der Institution', transcription of a fragmentary paper in the Gotthard Günther archival estate (Staatsbibliothek Berlin, manuscript department), accessible online: https://userpage.fu-berlin.de/~gerbrehm/GG_Institution.pdf [accessed 9 April 2024].
4. Maeda and Makino, 'A Pulse-Type Hardware Neuron Model with Beating, Bursting Excitation and Plateau Potential', p. 99.
5. See McLuhan and McLuhan, *Laws of Media*, pp. 36–38.
6. This expression is a technological analogy to 'the kind of theory which is an element in action': Horkheimer, 'Traditional and Critical Theory', p. 216.

objects instantly in analogy to the electric short circuit itself. A specific kind of critical media-philosophical contemplation arises from a techno-active co-presence that undermines the logocentric dichotomy between mind and body, *lógos* and matter or software and hardware, in favour of a dynamically entangled, ontologically interrelational tec-knowledge. Before this results in elaborate media-theoretical verbalization, this cognition takes place in the form of stenographic flashes of insight: techno*graffiti*.

The *transitive* approach of technical media criticism occurs both ways: There is human cognition of the technological—that is, material and techno-mathematical artefacts on the one side, and the appeal *from within* technological devices to the co-enactive mind on the other. As long as everyday tools are used practically, they are rarely consciously reflected as technical beings, unless in moments of malfunction when they disclose their proper idiosyncrasies. Transitive media philosophy practices a kind of media reasoning that has them 'ready-to-hand' not in a reduction to their instrumental function but rather in order to experience such assemblages as 'ready-to-knowledge' in the epistemic sense.[7] Transitive media philosophy is tracing the techno*lógos* from within the machine: be it by direct contact with hardware, or via symbol manipulation (software). The aim of such operations is not instrumental but ontological—in a transontological sense, though. It does not simply intransitively philosophize *about* technical media but allows for a *media thinking* that lets the technical event co-articulate itself.

The Micro-Tempor(e)ality of Transitive Media Philosophy

A paradox arises: How can a *media* experience be *immedia*te, if any material channel of signal transmission (quantum entanglement notwithstanding) inherently implies a delay and involves noise? Differing from subsequent metareflexive thought, *inductive* media reasoning comes close to the engineering sense of electromagnetic induction while at the same time staying close to 'induction' in traditional philosophical reasoning.[8] Such flashes of insight arise in direct contact with techn(o-mathemat)ical signal transduction and data processing. An electric circuit is not simply one of the many possible objects of such an investigation but becomes its very model: an immediate epistemic action-reaction in terms of a transient or 'flat' media ontology.[9]

7. See Harman, *Tool-Being*, p. 16.
8. On induction in the sense of Gilbert Simondon's technical ontology, see Hui, 'Deduktion, Induktion und Transduktion'.
9. For the definition of such a non-hierarchical ontology, see DeLanda, *Intensive Science*, pp. 46–47.

The computational equivalent to such an experience, where 'everything happens' within a fraction of a second 'in excess of the actually performed action and of its ascribed meaning',[10] was suggested by Donald E. Knuth on occasion of the 11th World Computer Congress in 1989: a thorough analysis of everything a computer does during one second of computation.[11] Media-artistic researcher David Gauthier actually monitored and listed the single-step Linux kernel events during one second of his computer's processing (between 12:30:37.178624843 and 12:30:38.163742036 on 4 September 2018) by application of a GNU debugger that allows for such an introspection on an extremely time-stretched scale.[12] Instant tec-knowledge is located at the threshold where such an implicit event horizon becomes epistemically conscious. A different, non-Platonic *lógos* articulates itself at that moment when 'the primacy of the affective',[13] as it is known from kinematographic and electronic image perception, is replaced by an accelerated media-epistemic cognition: Epistemic insight in real-time that bridges the micro-tempo*real* gap that usually opens between affect and cognition. Similar to Whitehead's processual ontology, transitive media philosophy instantly 'prehends' the technical event.[14] Almost co-originally with the medium itself, epistemic intuition becomes explicit insight within a tenth of a second, the time-critical span it takes for an implicit cognitive impulse to become knowledge that is present-at-mind.[15] It is this instant quality that characterizes transitive media philosophy in terms of time.

The techno-logical unconscious can indeed be revealed by transitive media attention. Such a 'tecgnostic'[16] attunement resonates with the technical tempo*real* event. Electronic flash and 'flash of thought' coincide here, literally short-circuiting the electronic device with the human neural mindset. The embodied mind has a privileged (though differential) affinity with physically implemented technology. Media recognition is entangled with its objects in techno-reality. Transitive media philosophizing is therefore practised from enacting media themselves. Media 'thinking', if it is not reduced to a mere

10. Massumi, 'The Autonomy of Affect', p. 90.
11. See the entry 'Knuth's Challenge', https://www.gaxun.net/commentary/knuth-challenge [accessed 29 August 2023].
12. As explained in: David Gauthier, To *Execute, Rewrite, and Debug: On the Construction and Deconstruction of Computation*, doctoral thesis (Faculty of Humanities, Amsterdam University, 2020).
13. Massumi, 'The Autonomy of Affect', p. 84.
14. Whitehead, *Science and the Modern World*, p. 69.
15. On this perceptual time gap see Canales, *A Tenth of a Second.*
16. A neologism borrowed from Leiner (alias QRT), *Tekknologic, Tekknowledge, Tekgnosis.*

metaphor, can only be derived from their actual operations. In contrast to the functional concept of explainable technology (the engineering or computer science approach) and to some kind of rather intuitive 'hands-on' media-artistic research, the transitive media philosopher is provided with an explicit epistemic *bias* (in the rigid understanding of media archaeology). In medical terms, *diathesis* names the affective predisposition of a body towards certain diseases. By analogy, transitive media philosophy requires a cognitive awareness that diathetically reflects its material and processual medium reference in real-time.

Catching the Media-Epistemic Momentum: From 'Transitive' to 'Transient' Media Philosophy

Affect and reason, immediate phenomena and distanced reflection usually exclude each other in conventional epistemology. This gap is tunnelled under by instant media philosophy, which occurs in real-time where a different, non-phenomenological kind of intellectual affect occurs. Sparks of insight are struck from the direct coupling of human and non-human reasoning (be it direct material or algorithmically coded hardware). A technical objectification of the mind can be experienced and recognized more immediately than from a belated, textually filtered distance. It comes close to that kind of improvisation in musical performance that intuitively is aware of its own music-theoretical premises. The title and subject of media artist Joost Rekveld's PhD thesis *Dialogues with Machines* 'refers to a back-and-forth between devising and observing' technological agencies. Implicit in this process is 'a relative absence of hierarchy between verbal thinking and making, and between the maker and the made'.[17] The chapter 'Liberate the Machines!' provides the theoretical and historical context to the making of Rekveld's corresponding film *Mechanisms Common to Disparate Phenomena; #59*: a film that focuses on the experiences *with* and *of* analog computer graphics shared by humans *and* electronic circuits. Transitive media philosophy is the epistemic equivalence of such an awareness. It lets tec-knowledge be liberated from its familiar anthropomorphism *by* the machine.

A media-cultural wonder occurs when, some time after the strange encounter of a singer with an electronic recording device, a poetic voice emanates from the cold wire spool in bodiless technical replay.[18] The moment when a

17. Rekveld, *Dialogues with Machines*, 'Abstract'. On Joost Rekveld's research project *Dialogues with Machines* in the Media Arts Studio of the KASK Conservatorium (University College Ghent), see http://librarinth.joostrekveld.net/doku.php?id=start
18. This experience has been described in Lord, *The Singer of Tales*.

Figure 1.1: 'His Master's Voice'.

human speech signal (such as epic oral poetry) emanates from a circulating spool, the apparently 'dead medium'[19] of the magnetic wire recorder is revealed as a *nonhument* (a term for technical storage as cultural monument) for transient media philosophy. The horizon of philosophical reflection, cultural tradition and academic thoughts about 'bodiless voices',[20] is condensed into a flash-like but at the same time media-epistemically conscious, transitive *knowledge affect* (no oxymoron here). Transitive media philosophy, in its media-archaeological bias, instantly replaces the auratic moment with a rational analysis of the techno-logical momentum. Its enquiry is driven by scientific curiosity and critical enlightenment and does not remain stuck in the affective experience. Even the seminal portrait of a dog listening to his master's voice from the funnel of a gramophone becomes a question of technical intelligence at that moment. The occurrence of a 'human' voice from

19. See Bruce Sterling, *The Life and Death of Media* (lecture at the Sixth International Symposium on Electronic Art ISEA '95, Montreal, September 1995), http://www.alamut.com/subj/artiface/deadMedia/dM_Address.html
20. As declared for radio plays by Kolb, *Das Horoskop des Hörspiels*.

a most inhuman recording device by means of mechanical (gramophone) or electromagnetic (record player) induction is a techno-traumatic irritation of logocentric presence,[21] and calls for critical media-philosophical awareness already at the transitive moment when such acoustic signals address the hearing apparatus.

Similar to the notion of transitivity in this volume, in electronics, ephemeral phenomena, especially in the form of a very brief surge of an electric current, have become a medium-like technical term: *transients*. In acoustics this names a 'non-repeating signal in an audio waveform that occurs very quickly', and computer programming knows transients as a 'module that generally remains in memory only for a short time'.[22] In terms of media philosophy, a transient is a flash of thought that arises in moments of media immediacy. It does not derive from intransitive musing *about* media but from operative experience below the time-critical threshold that separates hermeneutic insight (in the humanities) from scientific measuring.[23] Such an epistemic momentum can effectively be derived only from a close contact with the medium 'in being'—not from a passive reading or decoding of its electronic diagram or computational source code. The technological mode of existence never 'pre-exists' its medium event.[24] While media-theoretical distance allows for delayed reflection, it misses the risk of epistemic irritation (even 'veto') that may be articulated from the technical artefact when it is directly experienced in direct contact, as seen in the transients in communication engineering. Such electric shocks occur as momentary bursts in the physical world. Highly condensed microchips are especially vulnerable to transients. Beyond their functional disruptions, transients deserve to be instantly contemplated as irruptions of the energetically real into the symbolical order of computation. By analogy, transitive media philosophy arises from short circuits between technological knowledge and the mate*real* reality. Instead of a metadiscursive (deductive) approach to the phenomena of material aesthetics, transitive (epistemically inductive) argumentation arises from within a mental co-enactment of material techniques. Such inferences usually recede to imperceptible momenta and escape the analytic mind—unless that mind (be it human, or 'artificial') has been trained to recognize them as technological intelligence.

21. For such an analysis see Dolar, *A Voice and Nothing More*.
22. https://en.wiktionary.org/wiki/transient [accessed 5 November 2023].
23. As analysed by Siegert, 'Das Leben zählt nicht', pp. 161–82.
24. A media-specific variation of Manning and Massumi, *Thought in the Act*, p. 11.

Media vs Medium [Philosophy]

In this present treatise the terms 'medium' and 'media' oscillate between a strictly physical and a technological sense. Communication studies analyse the impact of media on human individuals and collectives from a data-based, statistically empirical distance. A medium archaeology instead investigates their infrastructural techno-logical materiality and tempor(e)ality in direct technical contact for instant epistemic reflection, for instance in terms of a *matterphorology*. Whereas the linguistic *metaphor* is 'defined as the trope of meaning transfer and substitution by means of analogy, matterphor denotes the articulation of meaning in relation to matter, understood not as fixed entity, but as constantly shifting (-phoric) and thereby establishing entanglements and relationalities'. This necessarily results in a different way of doing media philosophy: not as a cognitively distanced thought exercise but confronting tec-knowledge in its 'many embodied and situated actualizations' instead.[25] That is what media engineering experiences every moment when an abstract circuit design is *actually* confronted, in its technical realization, with the resistances and affordances of matter and energy.

In contrast to a naturalist nostalgia for the agency and meaning of matter as a counter-reaction to a highly technologized and digitized world, transitive media philosophy traces the *lógos* of technically informed matter as a specific instantiation of a mind that can never be pure reason. Rather, in reverse: *Lógos* is not colonizing matter here in a patriarchal (literally logocentric) gesture but is instead understood as a function of matter's inherent desire to transcend itself as epistemic morphogenesis, driven by a techno-logical bias (or even intelligence) that still remains to be specified. This is not anthropomorphizing the hardware and software of technology but granting an epistemic (not just functional) agency to technology itself. Human interaction(s) with matter or energy were previously forced from the outside, with both being subjected to brute mechanical or subtle electronic manipulation. With the universal Turing machine, the information-processing paradigm as physical modelling, and digital signal processing, the physical world itself has become rather intransitive ('as if'). The *medium* that in communication engineering has been traditionally defined as a passive channel of signal propagation becomes *media-active* once it is techno-mathematically considered a fusion of algorithm and machine. The external relation to material, energetic and informational processes has been internalized in negentropically prefigurated microchips for calculating with(in) matter itself, on the border

25. Gandorfer and Ayub, 'Introduction: Matterphorical', note 23.

with quantum computing. In a more philosophical sense, the very term 'technology' is not reduced to external academic or engineering knowledge *about* technique any more. The term rather extends to technical thought itself: to an original coupling of cognition and technique as theoretical machinery in its own right. This is neither a natural philosophy of technology nor a reconciliation with cosmic aesthetics but a technocosmic understanding of the human–machine coupling itself.[26]

The 'Language' of Media [Philosophy]: Transitivity

The transitive media philosopher, even when reading a technical report rather than in direct confrontation with the machine, is charged and biased (in the engineering rather than discursive sense) with media-epistemic curiosity and immediately deciphers this code as evidence from which sparks of media-epistemic insight arise. Philosophical attention is paid to the technical details and immediately weaves them into fundamental ontological issues, instead of separating the technical description from more media-theoretical thoughts in the form of detached texts. Transitive media philosophy is partnered with a mode of writing technology where precise stepwise analysis of the artefact and its techno-mathematical description are interwoven with its media-epistemic surplus insights. Gilbert Simondon practised such a kind of *archaeography* in his minute differentiation of electronic vacuum tubes, from the diode to the triode to pentodes and beyond.[27] As long as the logical description of a mechanism and its circuit diagram are considered equivalent to the machine itself, this can take place in the symbolic regime, but this changes when the actual frictions of the material machine are taken into the epistemic account as well.[28] When such an attention is directly confronted with a technical artefact, the processual device is experienced as if it were already media-theoretically coded, implicitly. The technically *analog* device is then understood as material *analogue* to a media-theoretical textual expression. But technical devices are in *medium*-being once structured matter processes signals. This comes close to an extended definition of the 'verb' in linguistics, indicating a physical or mental action. '[W]hether or not it is used with a direct object'

---

26. On the notion of 'cosmotechnics', see Hui, *The Question Concerning Technology in China*.
27. Simondon, *On the Mode of Existence of Technical Objects*, Part I, Chapter I 'The Genesis of the Technical Object: The Process of Concretization', pp. 29–40.
28. See Ernst, 'On the Notion of Cybernetic Frictions and its Role in the Deployment of a Radical Media Archaeology', pp. 119–30.

defines its transitivity.[29] By analogy, media are assumed to be 'verbal' when medium-being is essentially defined by its operative signal processing. When the ontological question shifts to 'What can a medium do?' (as this book series explores), media philosophy itself is reminded of its processual core.

In less object- than *process*-oriented terms, transitive media philosophy deals with the structural transitivity and time-critical short circuits between the symbolical order and its technical mate*reality*. Leaving the quantum physical microcosm aside (though that will increasingly affect the techno-logical regime), it is not from brute physical matter itself that symbolic articulation occurs; rather, technical logics (such as a network of interconnected electromechanical relays, or the modulation of electromagnetic waves) are impressed upon matter or energy to become media-active agencies. This moment of transition certainly deserves philosophical attention and media-epistemic questioning. 'Having expressed our premises upon the machine, and their effect being recorded by the pointers, it only remains to read off the conclusion', Allan Marquand commented on an operation of his Logical Machine, which itself had been inspired by William Jevons's Logical Piano.[30] Is such a techno-logical articulation resulting from an external (negentropic, 'informational'[31]) impression that acts invasively upon the material system, or rather from its implicit techno-mathematics? Is such a mechanism slavishly simulating logical reasoning or is its materiality, once thereby in/formed (as with, for example, arbitrarily contaminated semiconductive silicon), in fact *a*ware of such operations, as a kind of hardware emulation of soft logic?

Articulation in/from/as Matter: Writing the Machine Archaeographically

Media philosophy, in its transitive understanding, differs from ontological abstraction in the Platonic tradition. A genuine media philosophy worthy of the name does not write intransitively *about* technical media but identifies their philosophical qualities and impacts *from within* its technological experience. When Shannon et al. wrote about speech signal transmission, their explicit 'philosophy' of pulse code modulation did not relate to metaphysics but to the engineering aspects (such as the signal-to-noise ratio) of

29. *Collins Advanced Learner's Dictionary*, https://www.collinsdictionary.com/de/worterbuch/englisch/transitivity [accessed 5 January 2024]
30. Marquand, 'A New Logical Machine', p. 305.
31. Peters, 'Entropie und Information', p. 166.

discrete vs continuous signal transfer in terms of analog-to-digital sampling.[32] Truly archaeographically,[33] the text does not simply write *about* electric speech transmission but remains close (i.e. transitive) to the actual medium event of telephony and radio waves. From a hermeneutic angle or a media-theoretical perspective, the immediate entanglement of technical report and media-epistemic conclusions may seem premature. But this kind of short circuit is exactly the kind of insight that takes place in the transition from preconscious 'prehension' to the proto-reasoning of technology: 'Apprehension without the "ap-" and with the emphasis on the new first syllable: prehension. It all begins with an uncognitive taking account of an as-yet-indeterminate apportioning out'.[34] But it requires turning that oscillating epistemic spark in the resonant human–machine circuit (the fast track of technological insight) into the 'writing techniques' of philosophical thought, in order to reveal its techno*lógos* explicitly. Not every technical action is *a priori* 'a mode of thought, already in the act'.[35] It rather becomes so by diagrammatic reasoning. The moment of transition in between is exactly where transitive media philosophy occurs. This usually happens at the cost of transforming immediate signal action into an intransitive metalanguage that replaces the technical being with its symbolic representation: textual media theory. Transitive media philosophy, on the contrary, is analogous to the following imperative: 'Don't just write *about* dance. As William Forsythe says, dance that thought around.'[36]

Roland Barthes has reminded of a linguistic *genus verbi* in the Indo-European languages family, in between active and passive voice, where the subject is not simply governing the action but is retro-affected by the event itself. This coincides with the difference between metalanguage and object-oriented language that is operational and transitively linked to its object. The woodcutter names the tree that he is felling: He speaks the tree, he does not simply speak about it.[37] Barthes's analysis ends with the plea for 'a reconciliation between [...] description and explanation, between object

32. As expressed in the title of Shannon, Pierce and Oliver, 'The Philosophy of PCM'.
33. On this neologism see Hiller and Höltgen (eds), Archäographien.
34. Whitehead, *Science and the Modern World*, p. 69, as paraphrased by Manning and Massumi, *Thought in the Act*, p. 23.
35. Manning and Massumi, *Thought in the Act*, p. vii.
36. Manning and Massumi, *Thought in the Act*, p. viii, referring to the performance *Woolf Phrase* by the Forsythe Company (2001), where the rhythm of Virginia Woolf's novel *Mrs Dalloway* has actually been choreographed. 'From its conception, the piece is about movement and language coming together *and in relay*.' Ibid., p. 31 (emphasis added).
37. Barthes, 'Myth Today', p. 146. See also Kaiser, 'Do You Speak Technology?'

and knowledge'.[38] By analogy, there can be a kind of media philosophy that does not project a predefined concept (theory) on an external technical practice, but maintains that kind of transitive relationship to material electronic circuit hard*wiring* even in symbolic *writing*. A little manipulation of a signifier makes all the difference here: from knowledge about technics to genuine *tec-knowledge*.

While a usual live transmission of a football game on television focuses on the visual perspective, a skilled radio reporter does not just describe the game from a distance but is explicitly metareflexive of the sportive moments while they occur. The immersive 'resonant interval' is defined by temporal simultaneity and audio-tactile immediacy.[39] By analogy, the traditional art of description (*ekphrasis* as known from art history) transforms into the question of how to actually write the machine transitively. Theoretician of kinematics Franz Reuleaux advised describing a machine not by abstraction but as verbal—that is, processual re-enactment of the mechanism as it operationally unfolds. His *Kinematik* in 1872 transcribed machine work into a formalized language, while Charles Babbage's 'mechanical notation' has been more ambivalent: Formal algebraic notation itself is already conceived as the machine.[40]

Lógos in/as Energy: The Electronic Circuit

A transitive philosophy of technical media directly addresses the key question of how *lógos* relates to (in-)formatized matter or energy. While human beings dramatically experience the gap between an ideally timeless mind vs an entropic (ageing) body, engineered devices make this dichotomy experimentable and knowable as an objectification. With such tools (or toys), reason is techno-physically coming into being: concrete logo-techniques as an alternative term for 'technology'. Once 'structurally programmable' matter[41] has been negentropically in/formed by electrical engineering into an electronic circuit, such a circuitry *has* knowledge. Because the technical entanglement of reason with hardware differs from human embodiment, such knowledge is of a unique kind. Its techno*lógos* knows about (and apprehends) different material articulations, such as frictions, options and connectivities, that cannot all be predicted in advance by the engineering mind. Transitive media

38. Barthes, 'Myth Today', p. 160.
39. McLuhan and Powers, *The Global Village*, pp. 22–23.
40. In the mechanistic sense of Turing's definition of the algorithm: Turing, 'On Computable Numbers', pp. 230–65.
41. The expression comes from Conrad, 'The Prize of Programmability', pp. 261–81.

philosophy, far from understanding such moments as exceptions from regular functions, pays epistemic attention to such articulations.

Experimenting with electricity has for a long time been directly linked to the laboratory 'desktop' apparatus. Its transitive experience was frequently illustrated by including the experimenter's hand and fingers, such as in the figure of a Morse telegraphy circuit in an 1882 edition of *The Electrician* journal.[42] Human engineering is explicitly 'in the loop' here within the iconic diagram of the technical event. Even after the abstraction of such elements into formal circuit symbols, some figurative references to its actual matter persist. The symbolic remains entangled with the real even if it recedes into (almost) intransitive coding. It is only with the 'scale-of-two circuit' in digital computing,[43] that the radical abstraction from matter and energy in favour of conceptual binary 'information' finally succeeds. The human–machine relation becomes intransitive and is shifted towards the design of metaphorical user interfaces where the auditory, visual or tactile signalling obscures the inner-technical event. The information measuring unit *bit* became almost autonomous from its concrete technical realization as an electromagnetic relay or a fully electronic flip-flop circuit.[44] It is not only mathematics that becomes machine here; logical reasoning itself is rather revealed *as* mechanism. This is most frequently addressed in engineering terms but may as well techno-philosophically be discussed as the core drama of technological being(s) in human culture: the intricate relation between the symbolic regime and the physically real. 'Boolean algebra exhibits complete abstraction of generality, and is therefore applicable to any system whose elements give it a structure which is of the same form as that of the basic notions of formal reasoning.'[45]

Transitive media epistemology questions such a logocentric idealization. It rather insists on the difference between (meta-)mathematical *computation* (the Turing machine model) and actual *computing* as its actual 'grounding' (in the double sense of electrical engineering and Heidegger's existential philosophy of being). Such a reasoning *of* technology (in its media-active meaning), though, can only be achieved when contact with the technical event is not completely lost in favour of intertextual theoretical abstraction. The functionally oriented immediacy of the iconic hand of the engineer in

42. See Dennhardt, *Die Flipflop-Legende und das Digitale*, p. 78 (Fig. 31).
43. Higinbotham, 'The Model 200 Pulse Counter', p. 710, as quoted in Dennhardt, *Die Flipflop-Legende und das Digitale*, p. 131.
44. Shannon, 'A Mathematical Theory of Communication', p. 380.
45. Craven, 'An Engineering Application of Boolean Algebra', p. 121.

early technical circuit drawings is analogous to the epistemically oriented experience in transitive media philosophy. At such a moment, what may look like a mere engineering application of algebraic reasoning in a rather passive material medium becomes a condition of possibility for cognitive revelations of techno*lógos* in its multiple forms of articulation. Functional technical manufacturing becomes transformed into an epistemic experience of technology: a knowledge matter with a philosophical surplus value in itself.

In logic and mathematics a binary relation is *transitive* in the case of 'if A is related to B and B is related to C, then A is related to C'.[46] Such functional mathematics can be understood as mathematical *mechanism*. While Marquand had presented an electromechanical relay version of symbolic logic, Peirce, in his 1886 letter to Marquand, actually sketched a fully electric circuit.[47] Peirce discovered the option of logically configurated electronics as basic epistemic operation. He inferentially designed the first circuit diagram of electrified logical reasoning. Diagrammatic iconicity, in the engineering case, may look purely symbolical but is always designed with its transitive (if not transcendent) 'becoming flesh' in mind. Only in its technical concretization does a circuit design (be it as conceptually perfect as possible) become 'sensitive' to the mate*real*.[48] This momentum frequently results in non-linear idiosyncrasies, frictions and path-dependent contingencies. Technical signal transduction—even if most rationally designed—always subverts the logical regime. Even the very term 'technology' is epistemically incomplete in that sense.

In/formation of matter or energy is not exclusively induced from the human side. When the electric circuit of a bistable multivibrator operates like a logical argument, it is co-enacted from within technology itself. For that reason the electromagnetic relay is a central protagonist in this treatise. As a guiding technical fossil, it incorporates the options and limits of a transitive approach to technology in an exemplary way.

In the present author's *Technológos in Being*, a kind of technical unconscious expressed itself concerning the electromagnetic relay, in the form of an autopoietic 'typo' of machine-writing. The quotation from Peirce's media-archaeologically path-breaking argument for the design of an electric circuit as an instantiation of logical reasoning, biased by a battery, reads: 'I think electricity would be the best think'—instead of correctly, thing—'to

46. https://en.wikipedia.org/wiki/Transitivity [accessed 5 December 2023].
47. For the sketch of a logical electric circuit by Charles S. Peirce, enacting Boolean algebra via binary-switching operations, see Peirce, 'Logical Machines', p. 632 (note 2).
48. For a 'circuit-sensitive' (German 'schaltungssensibel') media study see Dörfling, *Der Schwingkreis*, p. 359.

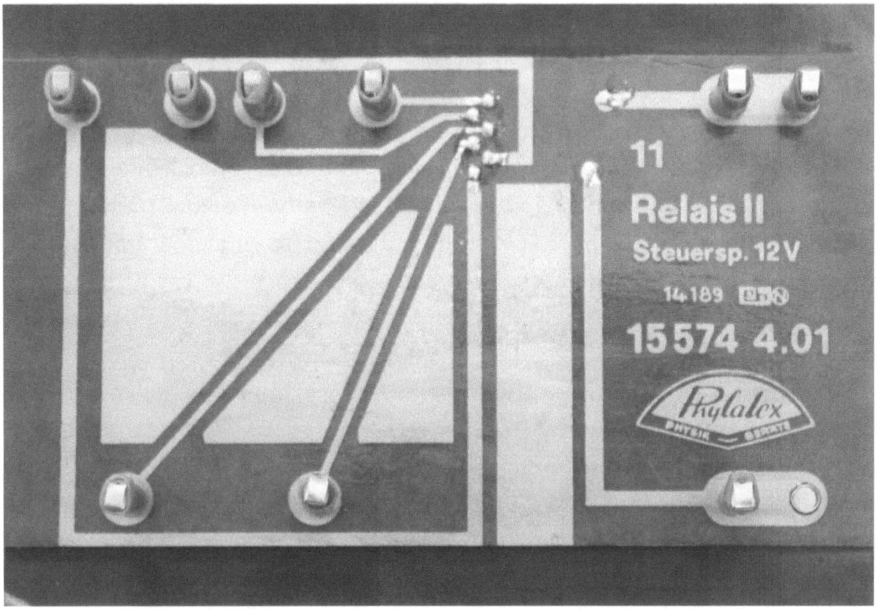

Figure 1.2: Electromagnetic relay as operative electric circuit.

rely on [...] where the circuit may be open or closed.'[49] What at first glance looks like a simple mechanization of multiplication and addition in logic, from the transient techno-philosophical perspective, becomes genuine media-active thinking itself. Such technological reasoning becomes experimental epistemology once media-archaeological circuit bending is instantly recognized in techno-philosophical terms, extending tinkering to 'thinkering'. First of all 'the tinkerer uses a "jumper" wire to connect any two points on the circuit board and thus temporarily short-circuits and rewires the device'.[50] Contingent signal events then occur just as from the non-linear connectivity of artificial neural nets: After sufficient training by big data, occasionally such processing sometimes results in a hardwired machine recognition of patterns that have been unknown to human insight before.

Transitive media philosophy, in catching hold of this moment, is not simply intuitive do-it-yourself (DIY) culture but is driven by the desire to experience a surplus of media-epistemic insight. Such an 'affective' media consciousness might appear to be an oxymoron, given the psycho-physiological definition of affect as a precognitive, subconscious mental event.

49. Peirce, 'Logical Machines', p. 632, as quoted in Ernst, *Technológos in Being*, p. 16.
50. Hertz and Parikka, 'Zombie Media: Circuit Bending Media Archaeology into an Art Method', p. 426.

Microtime-critical measuring media, such as the electroencephalograph, themselves revealed the belatedness of consciousness against precognitive cerebral impulse processing.[51] Manning and Massumi, though, insist that such a delay is no absence, lack or dead time but filled with potentialities and creative receptivity.[52] From the immediate, transitive, flash-like intersection of technical knowledge and epistemic recognition, media-philosophical insights into the core technological and inner-technical drama unfold. Precisely in that micro-temporal moment (or *epoché*), the 'media-epistemic affect' occurs. It is cognized by a tightly coupled human observer but induced by the mechanism itself.[53]

There is a specific kind of thoughts in this world that is generated once they originate technologically—that is, realized both *in* and *as* matter. To the attentive mind, in the encounter between epistemic analysis and a technological item, the co-original entanglement of cognitive and material knowledge becomes aware. The processual ('verbal') relation between human subject and technical object is not one-directional but becomes mutual. Its transitive media philosophizing is no immediate entanglement in the quantum physical sense,[54] but rather an already *lógos*-informed—that is, epistemically conscious—relation. Just like a transitive verb in the grammatical sense, think(er)ing media materially requires a technological object.

Can Media-Philosophical Reasoning Become Equivalent to Technological Circuitry Itself?

Transitive media philosophy is Janus-faced. It aims to instantly identify, to reveal and to make explicit articulations of a *lógos* that—at least as a heuristic hypothesis—can exclusively be expressed from technically configurated matter or energy—that is, in its medium specificity. The very term 'articulation' refers to logically in/formed matter, not simply to the physical 'real'. On the other hand, no processing machine can be reduced to an abstract philosophy of the symbolical form. In its actual mate*realization*, any cognitive representation becomes a physical signal. Its logic can only be expressed as *techno*lógos. In order to experience this ambivalence, experimental media archaeology, once it is intended as transitive 'thinkering', lets its media-theoretical distance

51. As described in Libet, 'Unconscious Cerebral Initiative and the Role of Conscious Will in Voluntary Action', pp. 529–39.
52. Manning and Massumi, *Thought in the Act.*
53. A diagnosis suggested in Angerer, *Nonconscious.*
54. The foundation of Barad's media-epistemic argumentation, *Meeting the Universe Halfway.*

collapse. Media-ontological questions are debated in the presence of their technical operations and practices—that is, with a focus on material processual form. Transitive media philosophy is not simply intuitive but closely related to media-theoretical concepts, but short-circuits them with their technical objects as co-articulating agencies. Observationally distant, metareflexive contemplation is thereby transformed into media-transitive reasoning. Instead of remaining *res extensa* in the Cartesian sense, technologies, in transitive media philosophy, are perceived as co-originary to the mind, and the dichotomy between thing and thinkers is reconsidered. On the one side, noumenology studies things as they are in-themselves, independent of or transcendentally beyond their immediate human sensation. On the other hand, Gaston Bachelard, in a paper on microphysics from 1931, proposed a 'phenomeno-technique' by which new phenomena are not simply found but invented: that is, 'thoroughly constructed'.[55] Technology infolds both worlds.

What appears at first sight like an accidental technical finding may transform into a more-than-technical insight in a highly epistemically charged field of enquiry. One example is the discovery of the oscillatory property of neon-filled lamps (Osglim lamps) that could then be applied to generate low audio frequencies (as with the choice of a glow lamp as tilting oscillator in the electroacoustic Trautonium prototype[56]) or high-frequency radio signalling. After a demonstration of this electronic quality, F.E. Smith remarked: 'Anson discovered this phenomenon "by accident", but such accidents only happen to observant persons.'[57] At such a moment, the inner-technical *lógos* and the electrotechnically informed mind coincide, creating sparks of insight that might be straightforwardly functional for engineers but may as well carry a media-philosophical surplus value. For epistemically curious observers charged with knowledge from the disciplines of science, technology, engineering and mathematics (STEM), this does not simply induce advanced technical practice; the same momentum can be turned into an epiSTEMic event by considering it as such. Transitive philosophizing, as *media* philosophy, requires continuous empirical orientation at actual machines, but in order to turn such techno-logical experience into media-epistemic insight, it requires *transcendental* (not transcendent) epistemic notions. While the concept of philosophical machines (machines for thinking, or thinking machines?[58])

55. Bachelard, *Études*, pp. 18–19.
56. Dörfling, *Der Schwingkreis*, p. 364.
57. Pearson and Anson, 'Demonstration of Some Electrical Properties of Neon-filled Lamps', p. 176.
58. See Laruelle, 'L'ordinateur transcendantale', p. 89.

largely ignores their actual technicity (hardware/energy), transitive media philosophy includes such engineering knowledge in its epistemic awareness. In such transitive media awareness, the philosophically familiar body–mind dichotomy is suspended by reciprocity, since the observing mind is dealing with a kind of matter/energy that has already been informed as technological object. Technical tools are thereby not simply understood as secondary functional externalizations or material objectifications of the mind,[59] but as a co-originary kind of reason that is unfolding in medium-specific ways.

59. A paraphrase of Gotthard Günther's characterization of the technical machine: 'Objektiv insofern, als es ein materielles Stück der Außenwelt ist, und Geist insoweit, als die Natur von allein keine Werkzeuge hervorbringt.' Günther, 'Die «zweite» Maschine', p. 220.

Chapter 2

From Within Technology: Signal Transduction

In analog circuitry, continuous transduction such as the conversion of an acoustic waveform into electric current is technically transitive. Its immediacy differs from time-discrete signal processing that conceptually involves numerical abstraction, intermediary calculation and short-term storage. For the media-epistemically tuned mind, the difference between transitive and intransitive technical action is reminiscent of Martin Heidegger's 'question concerning technology'. Heidegger differentiates between devices like the windmill that gently (transitively) adapts to natural forces on the one hand, and intransitive machine control (literally *cybernetic* 'governance') over nature on the other. That indeed makes all the difference between conventional techniques and more autonomous technology. Heidegger critically focuses on the techno-mathematical *attitude* towards nature; in that sense, the essence of technology is 'itself not technological'.[1] Just as Heidegger refuses a simply instrumental and anthropocentric definition of technology that is reduced to its human usage on the interface level, transitive media philosophy reveals a techno*lógos* from within the technically formatized material artefact itself.

The escalation from techniques to genuine technology is a media-philosophical concern. Archaeologist of science Michel Serres explicitly reserves the term *technology* 'for those kinds of artefacts that negotiate signs, that is: the logos', and contrasts them with *techniques* whose energetic range is 10^{16} times higher.[2] The related media-epistemic question is this: When does the transductive energy of electricity become informational electronics? It was not Edison's electric light bulb (the iconic original book cover illustration of McLuhan's *Understanding Media*) but its non-linear offspring, the thermionic vacuum tube (in its specification as triode), that allowed for techno-logical intelligence to be articulated from within the machine and which finally inaugurated the computational 'second machine' age.[3]

1. Heidegger, 'The Question Concerning Technology', p. 287.
2. Serres, 'Der Mensch ohne Fähigkeiten', pp. 194–95 (translation author's own).
3. As coined by Günther, 'Die "zweite" Maschine', pp. 179–203.

The Transitive Nature of Analog Media

Technically in/formed matter such as the phonographic engram (groove) differs from techno-logical (digital) encoding where the code does not transitively result from matter itself but is a function of an intransitive, external mapping of one alphabet of arbitrary symbols to another set. The digital computer, when looked at closely, is still analog electronics but dissimulates this technical truth since its signal processing is filtered by the symbolical code. One reason for preferring discrete coding against analog signal transmission techniques in wireless communication has been the improvement of its signal-to-noise ratio. The question of whether there can be tec-knowledge without noise is not left to engineering but is a media-philosophical concern. So-called analog media are transitive themselves by their technical definition, since in their operation the signals are instantiations of physical matter/energy itself. This became evident in the various interactions of light with photosensitive chemistry, especially in the early media-archaeological incubation phase of photography where the collodion wet plate process conditioned a different tempor(e)ality of optical aesthetics than its subsequent replacement by the dry plate procedure (glass plates with a photographic emulsion of silver halides suspended in gelatin).[4] Once the epistemic attention is directed to the lengthy process of developing a light-sensitive medium (including its smell of chemical fumes) rather than concealing it behind the iconography of the final image, the emergent pattern itself becomes a revelation of medium knowledge in terms of techno-organic (rather than mechanical or 'digital') *imaging*.[5] In phonographic recording, a similarly drastic transitive relation unfolds between the acoustic signal and its material inscription, from tinfoil over shellac to the vinyl record. The reproducibility of an analog sound recording (conditioning the industrial distribution of music that thereby became 'popular') depends on electro-chemical operations like the galvanization of the master recording. But substance itself is not yet technologically intelligent; 'matters of activity' become media-active rather by the intervention of arbitrary engineering.[6] Its technical reason, though, stays transient (rather than abstract) to the material event.

---

4. For a quick orientation see https://en.wikipedia.org/wiki/Collodion_process [accessed 24 February 2024]. Ideally transitive media philosophy goes even deeper into chemical details to reveal the epistemic delicacy of analog photography.
5. Such has been the aesthetics of Akosua Viktoria Adu-Sanyah's media-artistic installation *Corner Dry Lungs* at Zollart in Frankfurt (September 2024 to February 2025).
6. See the Cluster of Excellence *Matters of Activity: Image Space Material* at Humboldt University of Berlin, https://www.matters-of-activity.de/en [accessed 24 November 2024].

Shannon once defined the material channel for message transfer as 'merely the medium used to transmit the signal from transmitter to receiver. It may be a pair of wires, a coaxial cable, a band of radio frequencies, a beam of light, etc.'[7] By its intelligent coding, the signal is no longer decisively affected by the noise of the material channel. At Humboldt University of Berlin, the research project *Symbolic Material* changed perspective from a (apparently) passive notion of materiality, in favour of active matter that is able to 'generate symbolic processes from its own infrastructure'.[8] But such implicit affordances within brute matter/energy in itself can only be elicited by transforming them into tec-knowledge, whatever kind of biological, chemical or physical activities they might display otherwise. There is a specific kind of preformatism (or Aristotelian hylomorphism) in the technical engineering sense. In philosophical terms, the question remains whether techno*lógos* only emerges from already reasonably in/formed hardware, or whether such deformations are already prestructured by matter itself. Is such a determinism still technical or associated with another kind of reason? Transitive media philosophy traces such questions in the most immediate way. Electronic circuitry, down to present microchips, is not simply an external imposing of cultural reason upon matter. In the moment of transition between the symbolical operation and the physical real, matter is rather induced to reveal its inherent *lógos*. Neither can *lógos* articulate itself abstractly (as *langue*) without material embodiment (as *parole*), nor can matter articulate its reasonable potentialities without negentropic techno-logical intervention. It is this zone that transitive media philosophy identifies as technology in its infinite shades of insight.

Technology Becoming Intransitive? Digitization and/or Insight into 'Chaotic' Behaviour

One day, after the successful identification and repair of a mechatronical defect (the electromagnetic Hall effect sensor) in the ignition unit of an old-timer vehicle, the car mechanic confessed his relief: Such an archaic engine still allowed him to identify the error by direct transitive investigation of the machine. In contemporary software-governed motors an automated 'data report' by the analytic intervention of a computer is required instead. The issue of technical immediacy raises the question of how continuous analog *signal* transduction relates to *data* processing. Computation in discrete

7. Shannon, 'A Mathematical Theory of Communication', p. 2.

8. 'die in der Lage ist, aus ihrer inneren Struktur heraus symbolische Prozesse hervorzurufen': https://de.wikipedia.org/wiki/Exzellenzcluster_Matters_of_Activity._Image_Space_Material (translation author's own).

states is an arbitrary alphanumeric coding, a transcription of the physical (signal) into the notational regime (symbol). To what degree is the 'digital' a reductive abstraction against the material continuities that occur in micro-physical technical worlds? In 1937, the same year that Alan Turing published his seminal article on computable numbers, Edmund Husserl finalized his thoughts on the crisis of European sciences, arguing that algebraic mathematics had become oblivious of its more transitive 'eidetic' roots in agricultural geometry.[9] The mathematical intellect (Descartes's analytical geometry) has lost its sensual contact with the earthly ground of cultural techniques. *'Une fois évacuée la question de la mesure et partant de l'altérité de la matière physique, la correspondance entre l'écriture mathématique et le monde est exacte: le monde est une machine à états discrets, une machine à ré-écrire, simple transformation de suites de signes en suites de signes.'*[10] In other words, there has been an arbitrary decision taken to model the world discretely in order to make it accessible for computational approximation in the mathematical episteme. The arithmetization of geometry has become a central feature in digital computing while the analog computer operates in non-algorithmic modes. Whereas a directly electronic oscillogram is a transitive, indexical emanation of its technical configurations, the pixel image on a computer monitor is an indirect visualization of an otherwise symbolically organized machine. That makes all the difference between continuous signal transduction and discrete data processing.

In the years from 1954 to 1959, Herbert W. Franke's series of visually rhythmic *Oscillograms* (also called *Pendulum Oscillograms*) were not generated algorithmically in the digital sense. They were created with an analog computer instead, built by the Viennese physicist Franz Raimann in response to Franke's media-artistic demands. In a less aesthetic but scientific sense, as has recently been media-archaeologically remembered and re-enacted by artistic researcher Joost Rekveld,[11] on 27 November 1961 at Kyoto University in Japan, Yoshisuke Ueda discovered chaotic behaviour (the steady states of a non-linear system) during his experimentation with mathematically composed oscillograms. Such rhythmic patterns were not created as 'generative art' but were immediately coupled to the demands of scientific insight. Ueda

9. Husserl, *The Crisis of European Sciences and Transcendental Phenomenology*, p. 41. See also Friedrich Kittler, 'Phänomenologie versus Medienwissenschaft' [no year], http://hydra.humanities.uci.edu/kittler/istambul.html [accessed 8 November 2024]

10. Longo and Lassègue, 'Actualité de Turing', p. 231.

11. See Rekveld's brilliant piece of both artistic and practice-based media research, *Dialogues with Machines*.

had expected a regular oscillation after configuring the mathematical formula, but non-linear behaviour in the asynchronic phase (the 'chattered egg phenomenon') has been expressed by the graphic output of his analog computer. In terms of mathematical analysis, chaos appeared as a new phenomenon, 'but it has has always been around. There's nothing new about it—only people did not notice it'.[12] Ueda became aware of the epistemic event in its transient immediacy. But it has been detected *by* and revealed *as* techno*lógos* itself. The computer has become the dominant model for experiencing the ways in which the human mind models matter (*physis*).[13] A more materialist media archaeology instead insists on the active role of the physical medium itself as a condition of possibility (the Kantean *a priori*) for the techno-logical event to take place at all. Ueda's first assumption when observing chaotic attractors was 'that my analog computer'—that is, the evidence-providing technique (or toy) itself—'had gone bad'.[14] For the logocentric mathematician, it is the algebraic analysis that makes the electronic machine equivalent (or analogue) to the physical event, especially concerning differential equations for non-linear dynamics.[15] In the computer *simulation* approach to weather prediction, for example, '[t]he equations to be solved represent our best attempts to approximate the equations actually governing the atmosphere by equations which are compatible with present computer capabilities'.[16] Inductive techno-mathematicians, on the other hand, derive their conclusions from the 'impure' physical behaviour of computing itself,[17] once logical and algebraic reasoning has become materialized as electric circuits.

Towards a Renaissance of Analog Computing

Transitive Media (TM) is the trademark of a marketing agency located in Carlsbad, California, where 'solutions-oriented strategies are carefully executed by digital experts'. With the company's slogan 'we think deeply and technically', transitive media philosophy is in accordance.[18] But in contrast to the company's branding, it rather considers the digital as epistemic event.

---

12. Ueda, 'Strange Attractors and the Origin of Chaos', p. 27.
13. Krämer, 'Geist ohne Bewußtsein?', p. 105.
14. Ueda, 'Strange Attractors and the Origin of Chaos', p. 27.
15. See Steve Smale, 'Finding a Horseshoe on the Beaches of Rio', in Abraham and Ueda (eds), *The Chaos Avant-Garde*, pp. 7–22 (p. 8).
16. Lorenz, 'The Butterfly Effect', p. 92.
17. Mira, 'I. Gumowski and a Toulouse Research Group in the "Prehistoric" Times of Chaotic Dynamics', p. 190.
18. https://transitivemedia.com [accessed 24 March 2024].

In terms of computer science, the discrete, numerical and algorithmic approach to the physical world is conceptually intransitive and an approximation only. It gains its worldliness rather by the physicality of electronics itself. Binary coding maps signal transduction into data spheres that can be treated mathematically. Is it this media-ontological gap that separates discrete computation from apparently continuous physical and biochemical nature? Or is such a notion of a continuous, differentiable matter a metaphysical imaginary of the digital itself? The techno-mathematical alternative to discrete *computation* is continuous *computing* from within in/formed (not just symbolically formatted) matter itself. The wiring of electric voltage-biased units embodies algebraic equations by technological analogy, doing mathematics from within the mate*real*. From a Pythagorean perspective, this is no simple 'simulation of mathematical processes by electric events' or invasive imposture of reason upon matter,[19] but a transitive understanding of a matter that calculates itself implicitly as it becomes audible with the sound of breaking waves at the seashore. Transitive media understanding, in its inherent affinity with sonic reasoning, turns the electroacoustic modelling of such a perception into epistemic insight. Gottfried Wilhelm Leibniz qualified musical listening as an unconscious (*nesciens*) mathematical analysis that takes place within the mechanism of the human ear.[20] Such an analog computing performs a form of material self-processing. In terms of the anthropic principle,[21] such physical operations still require human reasoning to become mathematically conscious via computational technologization. In this sense the ancient Greek notion of a *lógos* in nature becomes media thinking.[22]

To answer such epistemic questions, a transient awareness of computing itself is mandatory—that is, an awareness of its technical essence that is co-present to its operational usage. The interface of a computer (the alphanumeric keyboard and optical monitor) usually hides its electronic insides. If the human–computer interaction takes place at an open, white-boxed computer (discarding all kind of enveloping or protective design), a co-present usage *and* a conscious inspection of the technical machinery becomes thinkable. By analogy, the transitive reading of a printed book may not simply

19. 'nach dem Prinzip einer Simulation mathematischer Prozesse durch elektrische Vorgänge': Franke and Jäger, *Apparative Kunst*, p. 86.
20. 'Musica est exercitium arithmeticae occultum nesciens se numerare animi': Gottfried Wilhelm Leibniz, as quoted in the epitaph of Enders, 'Mathematik ist Musik für den Verstand', p. 7.
21. See Carter, 'Large Number Coincidences and the Anthropic Principle in Cosmology', pp. 291–98.
22. See Tousseul, 'L'illusion de la pensée-machine', p. 198.

mean decoding its alphabetic formations in a distant mode but cutting up its relevant characters with scissors and then materially re-editing them, as has been poetically practised by William Burroughs both with paper and with magnetic tape. Such an operative reading does not simply add marginal handwritten marks or notes on the page but materially intervenes. Insight into the symbolic order of the printing mechanism itself thereby becomes philosophically feasible.

In terms of a book title (*Digital Materialities*) that reads like an oxymoron but addresses a techno-logical truth, 'the distinction between the digital and the material world becomes increasingly blurred', since they are 'not separate but entangled elements of the same processes, activities and intentionalities'.[23] Still, even in terms of such a mate*reality* of the digital there seems to remain (and to insist) a gap between the concept of binary (symbolical) information processing and real-world physical matters of activity. This ambivalence is no weakness in the analysis of the technical condition, but it is the driving epistemic force of its development due to the ambiguous nature of techno-logical mate*reality* itself. The question of whether this gap is unbridgeable has led to a renaissance of analog computers. Such alternative computing does mathematics not intransitively as literal transcription (the sampling and quantizing of physical signals into binary code). It rather stays transitive to its analytic object: hardwired operative diagrammatics as physical computing from within the material and energetic world (its signals) itself.

In a close microscopic observation that short-cuts media-theoretical distance, even the digital computer, as hardware, is analog electronics. A price has been paid for the increased miniaturization and acceleration of microchip performance: The conceptual and engineering distance that conditions binary-switching operations in order to sufficiently differentiate between 'zero' and 'one' tends to collapse. Its material substrates are increasingly becoming vulnerable to its electrical, material and chemical environment: '[C]omputational architectures are fundamentally shaped by silicon's material properties and the technical practices required to control its porosity at the nanoscale.'[24] This very porosity opens a sphere between transparency and the opaque. But this evidence has become intransitive, for human perception, through symbolic code and algorithms.

The classical mechanical or electronic analog computer, on the contrary, remains transparent as technology. Even apparently abstract (numerical) and time-discretely sampled software is not symbolically embedded within

23. From the synopsis of Pink et al. (eds), *Digital Materialities*, https://www.bloomsburycollections.com/monograph?docid=b-9781474295789
24. Berry, 'Porosity and Computation', note 4.

microprocessors as idealized informational *lógos* but physical: a continuous computing with the material technical elements, just like the 'patching' of complex sound patterns with real cables on a modular electroacoustic synthesizer. Not only was a 'new electrical understanding of sound' revealed by means of equivalent-circuit diagrams of electroacoustic machines in the 1930s,[25] but it allowed for a transitive epistemic *understanding media* as knowledge-in/formed, activated matter in reverse. The physical programming of musical synthesizers, as concrete 'patching' by cables, still follows the conceptualism of the musical score. But its real internal material action, with its analog (frictions) and digital (algorithmic) incompleteness, results in behaviour that asks for transitive reflection about the core agenda of (at least) occidental epistemology as such: the interrelation between mind and matter. Technically informed matter is always aware of (and even *knows*) its information.

Like the 'bit' as information unit in communication engineering, computational logic is both a conceptual abstraction *and* its actual hardwired circuitry. The relation between software and hardware becomes most transitive with so-called 'backdoors': code that is embedded inside the integrated circuit of computer chips. In everyday communication devices they allow for low-level control of functionality but can be misused to make their infrastructure collapse. 'Trojans' are no longer mere metaphors for disguised malware, but come back to the archaeological truth behind the artificial horse in Homer's epic *Iliad*: the earthquake god Poseidon who made the walls of Troy collapse.

What occurs during the 'calculations' in analog computing (derived from counting with pebbles, *calculi*), in a technically close reading, is no discrete mathematics at all, no symbolical mechanism. The mechanical or electronic device itself rather operates in the continuous mode. Analog computing embodies mathematics not as transcription (the sampling and quantifying of physical signals into binary code). It is instead transitive to its analytic object, performing mathematics as physical computing from within the material and energetic world itself. This enables immediate human–machine interaction as opposed to intransitive control of digital computation as rule-governed symbol manipulation. With the Turing machine model the numerical processing literally intervenes. In terms of synchronization, this makes all the difference between 'live' and delayed human–machine interaction. In creative programming this becomes evident in comparisons of the

25. Wittje, 'The Electrical Imagination', p. 40.

transitively modulated modular analog electroacoustic synthesizer vs digital programming environments like SuperCollider for real-time algorithmic audio synthesis.[26]

Media Epistemology of Analog-to-Digital Transduction

Philosopher of technology Gilbert Simondon conceived transduction in a rather generalized system-theoretical sense. As concrete techno-physically embodied operation, though, transducers sense signals and *transform* them (transitively) into another kind of signal (for instance, acoustic waves into electric pulses) or *transcribe* them (intransitively) into informational code that regulates the operation of the machine. In order to identify the essential operation (even metamorphosis) of analog signals becoming digital data in the electronic circuitry, the inner-technical world of an analog-to-digital transducer is the appropriate scene to investigate. More transitive than its abstract description, the analysis of a concrete technical configuration like the classic Fairchild Multimeter 7000 from the year 1969 reveals the idiosyncrasies of such a mechanism in its techno-epistemic delicacy.[27] Thereby, the intransitive description of a technical object (in the tradition of art-historical *ekphrasis*) turns into transitive media archaeography—that is, the conscious cognitive co-execution of the time-based signal event itself. Just as Gotthold Ephraim Lessing, in his 1766 *Laocoon* treatise, differentiated between the semiotic medium specificities of painting vs poetry, the spatial configuration of a hardwired electric circuit may be symbolically represented by a diagram but does not yet reveal the processual momentum of its material medium-becoming. The circuit unit that generates signal processing is analysed in as much detail as possible by a gradual re-enactment in the transitive thought experiment.[28] Its subsequent textual formulation occurs in a non-narrative style of media-epistemic plasticity.

When the Symbolic Regime Encounters Mate*reality*: The Transitive Spatiotemporal Momentum

The informational turn, as seen in works such as Wiener's *Cybernetics*, promised an epistemic tool for calculating the worlds of both machines and organisms as signal-processing systems that are analytically abstracted from their concrete

26. As discussed in Rohrhuber, 'Algorithmic Music and the Philosophy of Time', pp. 17–40.
27. Westphal, 'Vom Messen und Zählen', p. 102.
28. 'indem wir ihn Schritt für Schritt gedanklich aufbauen': Westphal, 'Vom Messen und Zählen', p. 106.

material or energetic realizations. Once conceived as information, signals 'can flow between carbon-based organic components and silicon-based electronic components to make protein and silicon operate in a single system'.[29] But information, in the radical media-archaeological sense, has no ontological but rather an operational quality. It is an abductive construct for communication engineering and technical reasoning. The software/hardware or code/matter divide collapses in favour of an equiprimordial (though not undifferentiated) form of technical existence. Transitive media philosophy addresses such core questions as how the symbolical regime (which is most familiar from computation) relates to the physical world. But it does so not in ontological abstraction but at the moment of experiencing technical events. While the inherited dichotomy, in occidental philosophy, has been between the discrete and the continuous (logical mind vs matter or energy), transitive media-philosophical thinking addresses exactly the in-between: the specificity of techno-logically informed matter. It is technology itself that both reveals and defers—but never suspends—the familiar dichotomy between matter and reason. As cultural technique, this encounter concretely takes place, for instance with material pencil on paper for the notation of a (therefore not merely 'symbolical') Turing machine. An electric circuitry, manually designed with graphical notation, is usually considered as merely symbolic. This representation may become media-archaeographically transitive when the graphic trace itself becomes conductive once an electric voltage is applied and becomes fully technological *in silicio*. The emphasis on always-already *physical computing* separates transitive media-scientific analysis from a pure (meta-)mathematics of *computation*.

Media-Archaeological Concretization: Relays as Techno-Logical 'Affordance' of the Symbolic Order

Media devices are not simply retroactive materializations but coextensive to their technological thoughts. They thereby incorporate technical potentialities of thought. They not only make matter 'thinkable' techno-logically but are actually doing thinking in medium-specific ways.[30] As a conceptual technical metaphor the relay, for instance, explains connectivity as discourse network both in theory and in practice: 'a relay from one practice to another', and 'a set of relays from one theoretical point to another'.[31] This artefactual agency

29. Hayles, *How We Became Posthuman*, p. 2.

30. As described by the editors of the present book series, *Exeter Concepts in Media Thinking* (Bernd Herzogenrath and Tomáš Dvořák): https://www.exeterpress.co.uk/collections/exeter-concepts-in-media-thinking [accessed 26 June 2025].

31. Deleuze and Foucault, 'Intellectuals and Power', p. 206.

becomes most concrete, though, in its actual enactment of information engineering. The relay as 'thinking' device is both theory *and* technology in its definition of the binary digit. Even if Shannon wrote an explicitly mathematical theory of communication, his vantage point has been applicable to engineering. His both technically concrete (transitive) and abstract (intransitive) approach articulates the techno-logical double-bind of 'thinkering'. The *bit* as quantitative, numerical (and therefore computable) unit where the choice of a logarithmic base corresponds to the choice of a unit for measuring information only *conceptually* allows abstraction from the technical physicality of matter and energy. Two stable positions, such as in the relay of a flip-flop circuit, can store one bit of information. But in its actual implementation that provides it with all its relevance in digital media culture, technology does *not* ignore the spatio-temporal and differential *in-between* of alternative switching states. Their critical cut is an energetic incision once it is enacted. Be it with pencil on paper as symbolical mechanism, or as actual machine: the informational transcendence of matter itself is a material act and thereby stays transitive. Binary calculation becomes a function of its electronic elements: 'Parameters of engineering importance such as time, bandwidth, number or relays, etc., tend to carry linearly with the logarithm of the number of possibilities.' Adding one relay to a group doubles the number of their possible states. From the technical mate*reality* of actual computing (rather than abstract computation) this technical operation 'is mathematically more suitable'.[32]

Such a technology is not simply an engineering extension of logical reasoning. While Alan Turing was mechanizing mathematical operations (the algorithm), his contemporaries Konrad Zuse and Claude Shannon treated technical elements such as binary relays as transitive mathematics. To operate his conditional calculus (German *Bedingungskombinatorik*), Zuse derived his computing from engineering practice. By using the symbolic notation of electromagnetic relays as a basis, he could design the circuits on paper 'without having to construct them until the design was complete'.[33]

From an engineering point of view instead of metamathematical deduction, Zuse privileged binary over decimal digits. This decision derived from the truly techno-logical affordance of the simplest possible technical unit: the electrical switch, 'which is either on or off'.[34] As an engineering *and* mathematics double talent, Shannon entangled *both* regimes: technical materiality and logical reasoning. Computational 'thought', as operative diagrammatics, does not simply

32. Shannon, 'The Mathematical Theory of Communication', p. 32.
33. Ceruzzi, 'The Early Computers of Konrad Zuse', p. 246.
34. Ceruzzi, 'The Early Computers of Konrad Zuse', p. 244.

equal electric switching by analogy, but becomes identical with it. In his master thesis, Shannon mapped propositional logic (Boolean algebra) upon networks of switching relays as a transitive relation.[35] While the contemporaneous Turing machine was a mechanization of mathematics, Shannon's approach mathematized the machine itself. Cognitive logic becomes enactive (in the cognitive science sense) once the mind is coupled to electrical engineering, in analogy to the tight, transitive coupling of brain and machine by the brain–computer interface (BCI). In the machine world, an implicit techno*lógos* indeed longs to be explicitly revealed by media-philosophical transitive reasoning: 'There is a sense in which all mechanical phenomena are expressive of logical relations. A typewriter contains hundreds of working parts that can be considered expressions of "and," "or," "if, then," and other logical relations.'[36] Modular material systems such as Charles Babbage's proto-computer *Analytical Engine* had already suggested the option of programmable hardware. But only an electronic computer is available for fully 'structural programmability'.[37]

Through the analysis of a set of physical events mathematically, matter becomes articulate and thereby computable. But can a mathematical approximation of natural phenomena ever become transitive to the actual event, or will a gap of intransitivity always insist? It takes already culturally in/formed and technically logified matter to make such worlds articulate their own symbolics. Once the single elements of an electronic circuit are treated as embodiments and equivalence of equations, in an emphatic understanding, such an algebraic treatment and physical informatization makes the symbolical logic of informed matter (such as a network of relays) speak itself in binary *articulation*. Electric trigger relays become seen as a materialized concretization of logical reasoning. Symbolic binary logic can be enacted as a physical event by means of an actual technical unit. The telephone-switching networks of electromechanical relays, in 1835, were not invented for logical reasoning at all but as a connective communication device. In techno-logical latency the implicit option of being applicable for binary logic and digital computing only became explicit by human reasoning. Transitive media philosophy locates such scenes where reasoning *lógos* and informatizable matter interfere: be it functional operations, or in the moment of unintended frictions. What is symbolically expressed in the alphanumeric code in 'digital' media actually exists and acts as differences in electric voltage.[38]

35. Shannon, 'A Symbolic Analysis of Relay and Switching Circuits', pp. 713–23.
36. Gardner, *Logic Machines and Diagrams*, p. 116 (note 3).
37. Conrad, 'The Prize of Programmability', p. 264.
38. Kittler, 'There Is No Software', p. 150–51.

In the end, even what seems like a most intransitive media processuality (the human experience of the 'digital') turns out to be a seamless chaining of technical transitivity. Against voltages that ideally count as 'one', there is not even a real 'zero', but always an electric ground bias. Binary states are therefore more appropriately termed 'high' and 'low' in communication engineering. Even what is metaphysically termed an 'absence' (the binary '0') is electrophysically present here. Only the logocentrism of numerical counting abstracts voltage levels into a binary alphabet. A 'bit' as information unit exists only conceptually. It becomes operative with the electromagnetic relay to perform the opening or closing of logical gates. *Lógos* is not simply materialized here but exists in this double-bind. In contrast to the embodiment concept in cognitive science, techno-logical implementations are controlled materializations of *and* as the reasoning mind—that is, as the venue where *lógos* and matter interfere. It is on this stage that the media drama unfolds that is recognized by transitive media philosophy on the fly.

In terms of a technical grammatology, the ostensibly immaterial Platonic *lógos*, in its withdrawal, is revealed by its material and/or energetic trace. An abstractly formulated algorithm, from a logocentric perspective in the ancient philosophical tradition, seems autonomous from its various concrete software implementations in computing machines. But concretely recognized as a stepwise procedure, it already has to become, or rather *is*, a machine that depends on a material stage (be it the calculating brain, or pencil on paper) to take place as such. Computation is always intertwined with the physical event: its transitive computing. In electronic terms this means signals that are wired within circuitry. Computational thought, here, does not simply equal but becomes symbiotically identical with electric switching as operative diagrammatics.[39] Transitive media epistemology therefore does not separate the technical signal from its theoretical significance. The techno-philosophical observer of binary-switching events does not simply re-enact its operations mentally but immediately expresses its epistemic surplus in terms of its techno*lógos*.

Concerning 'the Digital': Media-Epistemic Sparks from the 'Flip-Flop'

Volume 6 of the Berliner Programm einer Medienwissenschaft (Berlin Program of a Media Science) book series, Robert Dennhardt's *Die Flipflog Legende und das Digitale*, explicitly integrates the close analysis of a core

39. For an exemplary discussion of this claim, see Koopman, 'Microcodes versus Hard-Wired Control', pp. 235–42.

digital device into a historiographical genealogy of technical knowledge: the 'flip-flop myth'. Such an approach that does not smooth ruptures and frictions is media-archaeologically transitive in a double sense. It does not begin with a 'first' fully electronic digital computer but actually ends there (with the British Colossus and the US ENIAC/EDVAC in the early and mid-1940s). Its focus is on the transformations of the first Eccles–Jordan trigger relay from radio engineering into the core operator of binary computing. A conventional history of such technologies tends to privilege evolutionary linearities against the evidence of their discontinuous material embodiments. The transitive analysis rather stays close to the actual concretizations of the otherwise transcendent and abstracted signifier of 'the digital'. In this case, that is the mono-stabile trigger relay and the bistable flip-flop as the archaic electronics of digital computing. While another pre-history of the digital would get media-historiographically lost in an archival or bibliographical labyrinth of textual quotes, Dennhardt's analysis rather argues along the material basis of binary engineering and its diagrammatic modelling as circuit design. This comes close to the archaeological definition of 'pre-history' as non-scriptural epoch. With the technical apparatus and symbolic circuit diagrams as the protagonists,[40] digital techno*lógos* is identified in the field that unfolds between *termini technici* (*lógos* in documents) and material beings (*techné* as monuments). There is no digitality as such but only its concrete techno-logical unfolding *in being*,[41] be this the electromagnetic contact breaker or the electronic vacuum tube. The electromechanic relay, as known from discretely coded Morse telegraphy, actually anticipated binary information processing *avant la lettre*, as much as the early telegraph 'code books' established a techno*lógos* of word processing in between human and machine. The 'epistemic risk'[42] of speaking or writing just *about* the digital is counteracted by a transitive investigation of its actual implementations.[43] The abstract 'digital' (and the asymmetry within the techno-logical itself) becomes more transitive through a close technical observation of its core electronic devices, such as the binary flip-flop. Norbert Wiener criticized any ontologically sharp

40. Dennhardt, *Die Flipflop-Legende und das Digitale*, p. 8.

41. Dennhardt, *Die Flipflop-Legende und das Digitale*, p. 15.

42. 'das epistemisch Gefahrvolle des Sprechens in und über [...] das Begriffspaar analog-digital': Dennhardt, *Die Flipflop-Legende und das Digitale*, p. 34.

43. For an analysis that stays both philologically and linguistically close to the code *and* technically close to the machine, thereby swinging between intransitive (discourse-oriented) contextualizing of the device and its transitive (techno-epistemically entangled) evaluation, see chapter 5, 'Technogenesis in Action: Telegraph Code Books and the Place of the Human', in Hayles, *How We Think*, pp. 123–70.

distinction between 'the digital' and 'the analogic' already in its technical incubation phase: 'Every digital device is really an analogical device' that distinguishes a region of attraction by abstract coding rather than by a direct transitive measurement; 'a certain time of non-reality pushed far enough will make any device digital'.[44] It is only the cognitive representation that leads to an intransitive concept of the 'bit' that tends to ignore its techno-material tempor(e)ality. Electromagnetic relays in electronic circuits for binary 0/1 decisions switch in fractions of a second—that is, in the sublime, subliminal, affective time regime of human perception.[45] This ultra-short moment may be transitively addressed by kind of a subsymbolic media cognition that asks how the abstract symbols '0' and '1' as the binary alphabet of digital communication, or the numerical weights of artificial neurons in neural machine learning and AI, relate to their concrete embodiment as electromagnetic relays, transistors or chemically dotted silicon microchips. 'Information is information, not matter or energy.'[46] In epistemic dimensions, this is no mere engineering challenge but the recurrence of a core concern of occidental philosophy and religion. How does the rational human mind relate to the material physical world: dialectically, as entanglement, as embodiment or as technical instantiation? Any information still requires a material substrate that calls to be (techno-)logically in/formed. Even the most abstract digital 'bit', when observed closely, maintains such a transitive relation to the material world.

In the language of electrical engineering, Robert Dennhardt's media-historiographical narrative is continuously 'grounded' by non-textual circuit diagrams that are no simple illustrations. While Bruno Latour's actor–network theory and science and technology studies tend to extend the technical dispositif to non-technical actors like institutions or even to the 'social' as such, Dennhardt's analysis stays close to, and derives media-epistemic sparks from, the technical event that may be symbolically represented by an oscilloscopic time diagram.[47] The actually *operative* diagram of the binary digit as techno-logical being, though, is the binary trigger relay itself. The technical agency of the 'digital' procedure becomes concrete here. That is why, in Humboldt University of Berlin's Media Archaeological Fundus, a copy of Dennhardt's book has been placed in close proximity with a contemporary electronic realization of the Eccles–Jordan trigger relay.

---

44. Quoted here from: Pias (ed.), *Cybernetics / Kybernetik*, p. 158.
45. As discussed in Angerer, Ott and Bösel (eds), *Timing of Affect*.
46. Wiener, 'Computing Machines and the Nervous System', p. 132.
47. Dennhardt, *Die Flipflop-Legende und das Digitale*, p. 92 (Fig. 43b).

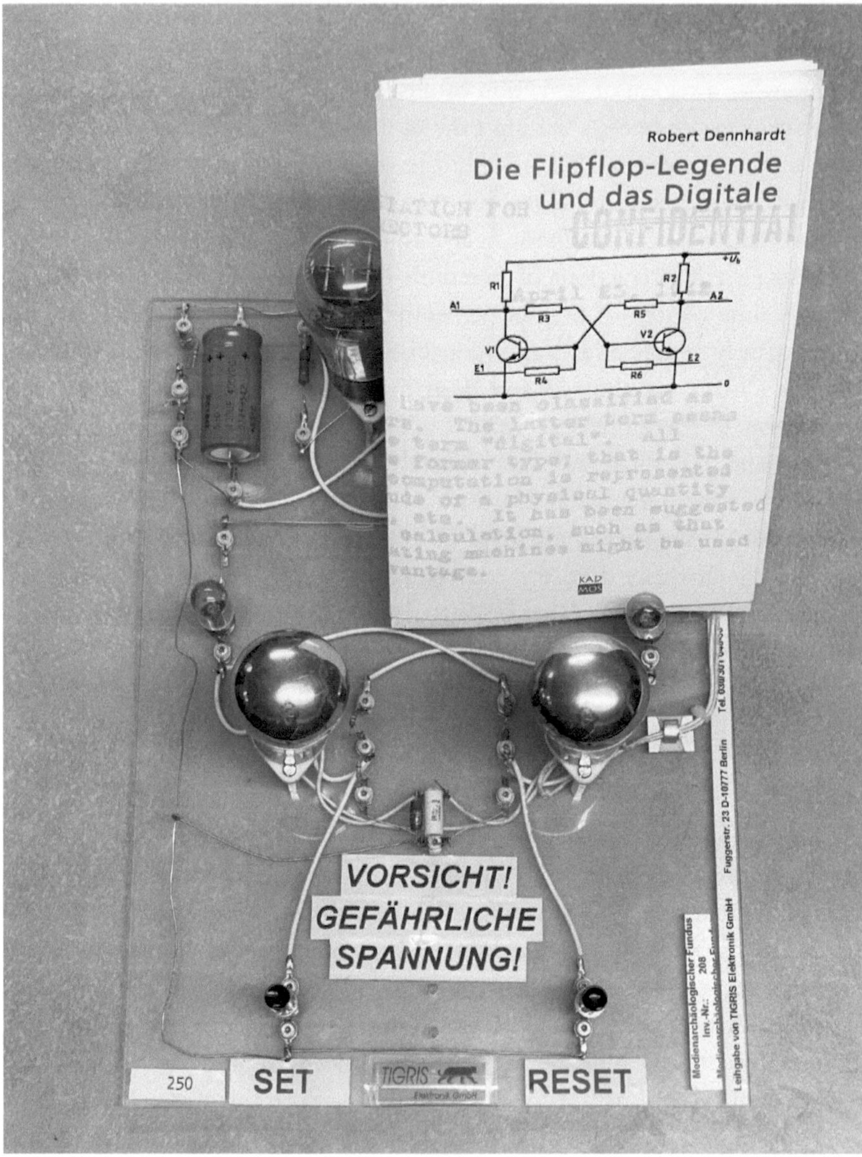

Figure 2.1: Symbolical and material presence of the trigger relay as technological object in the Media Archaeological Fundus.

What differentiates the digital computer from previous thermodynamic machines is its essence as a theory-born mechanism. Hardware enters, though, into this software thought experiment by the coincidence of metamathematical reasoning (the *Entscheidungsproblem*) with a techno-material affordance that as such evokes binary signal processing: the trigger relay. The 'origin' (*arché*) of the electric-digital is no logocentric abstraction or historiographically determined discourse, but the interruptive contact as originary apparatus (German *Ursprungsapparatur*[48]) that unfolds, varies and short-cuts across techno-historical time.

'The great thing about logic gates is that they are about less logic than electronics. Cognitive science doesn't get this, which is more than an affordance, but the schema [...] of digital thought itself.'[49] The material concretization of Boolean truth tables into electrified matter is not simply a mechanization of philosophical logical reasoning (such as a syllogism), but in terms of hylomorphism requires medium adequacy. A symbolic form hereby takes place *in*, or more radically *as* (in)formatizable matter. Operative *lógos*, in order to become media-operational in time-intensive intervals of movement or phases, roots in counting as material action.[50] The time integral of condensers, cross-related circuited transistors (flip-flops), time-giving electronic counters and differential amplifiers all force the circuitry to change from undefined states (their natural 'analog' condition) to either/or decisions that then become functional for binary 0/1 calculation within infinitesimally short time windows. Here Wiener's techno-poetic formulation of a 'time of non-reality' that for fractions of a second occurs between alternative switching states literally *counts* in electronic devices like the vacuum tube.[51] Its actual decision is truly random when it immediately derives from the inevitable thermic noise. Binary reasoning is challenged by its material condition and tentatively mastered by electrotechnical *lógos*. The electronic triode does not simply amplify a flow of electrons but allows a strong current to govern by application of a small one at an additional grid. All of a sudden, the energetic device transforms into a logical unit. Such applied cybernetics operates with inherent qualities of matter and energy itself, eliciting its implicit techno*lógos*.

Transient media philosophy makes this momentum explicit in its concern for the core essence of technology. The difference between cultural techniques of opening and closure (such as in architectural doors) and their escalation in technologies is a crucial media-philosophical concern.[52] Electromechanical relays, beyond their metaphorical discursive applications, become 'grounded'

48. Dennhardt, *Die Flipflop-Legende und das Digitale*, p. 42.
49. Electronic communication by Scott Lash, 23 March 2019.
50. Aristotle, *Physics*, IV, § 218–19.
51. Wiener's notion was proposed during the post-World War II Macy Conferences on cybernetics in New York. See Pias (ed.), *Cybernetics*, pp. 158–159.
52. See Siegert, *Cultural Techniques*.

here not only in terms of engineering but epistemically too. When mate*reality* can thereby be interpreted as discrete articulation, the symbolical is inherently embodied in the electrophysical real itself. This can be expressed in analogy to Turing: If the chemical basis of biological morphogenesis can be *analysed* by mathematical analysis, in return it can be assumed as discretely *synthesizing lógos*.[53] In binary signal transduction as information processing, the gate is a unit for deciding in logical or numerical calculation. The electromagnetic or fully electronic relay thereby enables the embedding of the intransitive symbolic order (logic) into the transitive physical real of engineering. The core operation of the binary-switching relay techno-logically thereby not only performs bit storage and processing. In immediate transitive awareness, it embodies an epistemic event as well.

Since Shannon demonstrated that Boolean logic could be mapped upon the same types of problems for which Babbage had designed his mechanical notation of computational machines, thinking the machine mathematically has been made possible. It was his techno-logical double qualification in engineering as well as in mathematics (sometimes asymmetrically conflicting) that made Shannon media-epistemically receptive for the electromagnetic relay to enact Boolean logic in binary steps. The electromechanical relay had been known from telephone network switching. In 1919, a fully electronic flip-flop circuit was described, *avant la lettre*, in a brief paper.[54] In unknowing media-archaeological anticipation of digital electronic computing, the authors describe their invention as a method of magnifying signals in electrical circuits for use in telegraphy and telephony. Implicitly, though, their circuitry already embodied the *lógos* of two stable states.

It was the material 'affordance' of this mechanism,[55] its techno*lógos*, rather than pure reasoning, that suggested the preference of the binary mode for actual computing. Leibniz, in his design for a virtual dyadic calculator, had already decided that the mechanization of computation with 0 and 1 was preferable to the decimal system. In the integrated microchips for high-frequency computing today, the logical design for the placement of transistors actually coincides with its technical materialization in silicon as techno-logical corollary. Algorithmic reasoning is not only applied to, but is

53. Turing, 'The Chemical Basis of Morphogenesis', *Philosophical Transactions of the Royal Society of London*, 237, no. 641 (1952), pp. 37–72.

54. Eccles and Jordan, 'A Trigger Relay Utilising Three-Electrode Thermionic Vacuum Tubes', pp. 298–300.

55. To make use here of a term defined by Gibson, 'The Theory of Affordances', pp. 67–82.

also evoked by, the design of switching circuits. Shannon experienced the *neural thinking* of such relays when the electric mouse 'Theseus' in his Maze-Solving Machine (an early instantiation of machine learning) got 'neurotically' lost in the loops of the labyrinth.[56] And secret communication research at the Bell Laboratories was oriented at the (literal, in this case) techno*lógos* of the Vocoder as voice encryption device. 'By using binary (on-off) PCM, a high quality signal can be obtained under conditions of noise and interference so bad that it is just possible to recognize the presence of each pulse'.[57] This logic had already been implicitly embodied in the electromagnetic relay as telegraphic 'repeater-regenerator'. Discretizing and enhancing a voice signal then becomes a most explicit form of technical *articulation*.

In radar detection too, the physicality of the electromagnetic pulse offered itself to be mapped upon discrete numerical computing,[58] while at the same time reminding that technological reason is irrefutably physically embedded information. In spite of information aesthetics' ambition to be ideally liberated from material or corporeal restrictions in the computable sphere, techno-logy as logic and mathematics that actually come into operational being—that is, medium-being—is always already (*en arché*) impure in its primary transitive entanglement with concrete instantiations. As actual technique—such as the trigger relay—*lógos* is always already entangled with the mate*real*. Information is physical, against any logocentric notion or Platonic number idealism of binary computing: 'In most instances a computer pushes information around in a manner that is independent of the exact data which are being handled, and is only a function of the physical circuit connections.'[59]

From Transitive to Intransitive Machine Writing: Programming

Actual computing is not simply 'the tangible proof, of a theorem in logic' as a computer program, but an electronic reality of its own. Rather than being a mere functional 'electronic realization',[60] algorithmic computation transforms into a medium existence when its *lógos* speaks as actual *parole* rather than immaterial *langue* (to make use of a core differentiation in de Saussure's linguistics). A computer programmer whose attention is charged with media-epistemic bias does not intransitively impose speculative philosophy onto the technological object (as occurs frequently in media theories). The

56. Shannon, 'Presentation of a Maze-Solving Machine', p. 175.
57. Shannon, Pierce and Oliver, 'The Philosophy of PCM', p. 154.
58. This thesis has been elaborted in Borbach, *Delay*.
59. Landauer, 'Irreversibility and Heat Generation in the Computing Process', p. 184.
60. Bolter, *Turing's Man*, p. 77.

programmer's sparks of insight are rather derived from the inner-technical event itself while it occurs. It is computer programming where techno-creative writing, in the digital age, becomes most concrete, such as in the differentiation between modular compile-time during programming and linear run-time (code execution). Such subtle differentiations do not only relate to programming as a task in the narrow functional sense. They have a much broader relevance for the understanding of current media culture, such as with the question of how tempor(e)alities in digital technologies depart from philosophically inherited phenomenological 'time' (as has been thematized from Augustine to Husserl). Some computer-programming languages are not *compiled* from written alphanumeric text to machine code but *interpreted*, 'meaning that the process of translation into executable machine code happens immediately, merging writing and execution of software'.[61] *Writing in the medium* (the grammatical middle voice, as known from literature or poetry) becomes a mode of computer literacy. Binary numbers lose their purely symbolic character once they become recognized as a machine input: Media-transitively, they actually '*are* the bits that open and close the computer's circuits'.[62] With today's high-level programming languages, the act of *writing the computer* has changed from archaic hands-on transitive manipulation of bits as concrete electronic switches to intransitive practices in so-called object-oriented programming and its encapsulations of data beings as 'the principle of the black box, the hiding of complex process behind a simple name [...] from programmer and user alike'.[63] Procedural programming results in an enduring *dis*simulation of tec-knowledge in digital culture. Python as a preferred programming language for generative AI hides most of its subroutines within its vast 'libraries'. Even coding thereby becomes *in*transitive to the machinery.

The epistemic rupture between counting and writing (and subsequently reading) had occurred already with the substitution of letter codes for number codes in Assembly, turning programs into a 'language' that computer-linguistically stays *intransitive* to the machine. More or less direct coding, as known from the punched card in Jacquard's loom, or from automatic musical machines, is no simple romanticism for tangible computing against computational abstraction but points at the central scene of the technological drama: the confrontation of the symbolic (computation) with the real (computing). If the invention of software allowed for the symbolic manipulation of a real machine, just as language and other cultural techniques allowed humans

61. Evens, 'Object-Oriented Ontology', p. 96.
62. Evens, 'Object-Oriented Ontology', p. 91.
63. Evens, 'Object-Oriented Ontology', pp. 91–92.

to model the physical world symbolically, it is still the material media theatre where this drama takes place and unfolds in its irreducible materiality.

The grammatical 'medio-passive voice' becomes computational in reciprocal interaction with programmable machines. Can the linguistic concept of the 'emic' (i.e. an insider's perspective of a particular communication culture) be applied to an operative analysis of technologies 'from within' as well? Transitive media philosophy works in both directions: There is the human cognition of machines and mechanisms on the one hand. But on the other hand—to allow for a somewhat anthropomorphic expression—a techno-logical device, as a materially objectified form of reasoning, itself is inviting (even longing) for human cognition in order to become (self-)conscious. This gives an epistemic sense to the concept of an 'alien phenomenology'.[64] The techno*lógos* hypothesis heuristically assumes in/formed hardware itself *as* a media-specific kind of software. Matter then is not simply a passive physical resistance against complete idealized logification. Through *transitive programming* (in the analog computing sense) as in/formation, matter is seduced to articulate itself as *lógos* in latency. The programming of an electronic analog computer (just like the modular electroacoustic synthesizer) takes place via direct material (not simply symbolic) wiring of cables that actually creates a new material circuitry. Can there be such a transitive programming of digital computers as well, even if the symbolic code (the abstract algorithm) is meant to escape the materialities of the machine? The low-level symbolic programming language is transitive in so far as it stays close to the actual machine language (its op-codes) or direct material coding. Even the high-level programming language BASIC allows for direct intervention into the memory register through its 'PEEK' and 'POKE' commands. The writer thereby maintains a transparent relation to the mechanism. Such operations become media-philosophical concerns once such programming is not simply applied computation but becomes consciously experienced as a form of enquiry into the essence of computation itself.

Lógos Embodied in Microchips

The escalation from discrete transistor circuits (TTL) to integrated semiconducting devices (ICs) once made computational logic less tangible for immediate human intervention, in favour of a more abstract techno-mathematical diagrammatics. In times where technological intelligence itself designs the highly integrated circuits for electronic computation

64. As has been defined by Bogost, *Alien Phenomenology*.

on the symbolical (coding) level in advance, techno*lógos* precedes media materiality. *There is* software indeed when it becomes a hardware description language. Between the transitive manipulation of electronic circuitry and the physically intransitive block diagram reigns operative hardwiring. In network communication, a ping flood or even a ping of death attack results in a denial-of-service on the target side but mostly leaves the hardware still unharmed. When a massive data attack overheats a microprocessor chip in a smartphone device, though, this might finally lead to exploding its lithium accumulator. Even though high-level programming was conceived to abstract (even liberate) discrete coding from the concrete involvement with the machine, the symbolical code becomes invasive—that is, radically transitive to hardware again.

Information can indeed turn physical,[65] deconstructing Wiener's 1948 anti-materialist definition of information as neither matter nor energy. Computation, already when it becomes operative mathematics with pencil on paper, is embodied as graphite. As an alternative to the familiar silicon microchip, graphite, once flattened to graphene, can be grown on a silicon carbide substrate chip. Carbon is capable of forming many allotropes such as graphene that are structurally different forms of the same element, due to its valency.[66] Valency itself is the chemical equivalent to transitivity in linguistics: the binding of specific elements. Charles S. Peirce's notion of the diagram as 'existential graph' literally is a paper machine when its expression (its graphite pencil) becomes materially transitive.[67] When a working graphene semiconductor may be controlled to conduct or insulate electricity arbitrarily,[68] the logical or numerical (digital) operation applied to it as electric voltage still seems to remain an external symbolical initiative. Can matter be made to think by itself? While Régis Debray's mediology deals with the 'becoming-material' of symbolic forms,[69] a decidedly transitive media hypothesis deciphers this relation vice versa as well. A transitive media-philosophical premise is to think through and with the symbolic *and* the real apparatus: its mechanisms and machinery.[70]

---

65. As has been insisted in Landauer, 'Information Is Physical', pp. 23–29.
66. https://en.wikipedia.org/wiki/Allotropes_of_carbon
67. See Schäffner, 'Topologie der Medien', p. 87.
68. As described in Sparkes, 'First Working Graphene Semiconductor Could Lead to Faster Computers'.
69. Debray, *Media Manifestos*, p. 8.
70. A core assumption of the *Thinking Media* book series, edited by Patricia Pisters and Bernd Herzogenrath: https://www.bloomsbury.com/us/series/thinking-media [accessed 14 November 2025].

Chapter 3

Concrete Scenarios: Moments of Media-Epistemic Flashes

With Foucault: Typewriting Media Philosophy Transitively

Any alphanumeric symbol, typed on a computer keyboard, becomes a physical signal once it is electrically processed from its 'subface'.[1] The keyboard of a typewriter or the interface of a computing device leaves the human user mostly intransitive to its inner mechanism. Pressing a key on a computer keyboard is usually associated with some kind of symbolic meaning, as part of a word or a sentence, and forming longer textual sections. This action then gets displayed on computer screens, enabling humans to read it. From a media-archaeological viewpoint, though, the keyboard strings of characters are transformed into electro-physical signals that thereby momentarily lose all of their semantic reference. Within a physical computer, the 'digital' sign instead gains electro-physical indexicality that provides for an operative function and application inside the electronic circuitry. This introspection of the 'algorithmic sign' induces a more complex understanding of the relationship between encoded symbols and their physical manifestation.[2] The transitive media approach is not restricted to discursive relations between humans and machines but becomes aware of the knowledge values of non-discursive technical configurations too.

What then is the articulation of a writing machine? A letter, printed as a character on a typewriter keypad, according to Foucault's *Archaeology of Knowledge*, is no statement (*énoncé*), since this term does not refer 'to the material act of speaking (aloud or to oneself)' as *parole* nor to 'writing (by hand or typewriter)'. A technical speech act (techno*lógos*, literally) is rather an operation that is carried out by the machine itself 'in its emergence'.[3] Accordingly, 'the keyboard of a typewriter is not a statement; but the same

1. Nake, 'Das doppelte Bild', p. 47.
2. Nake, 'Das doppelte Bild', p. 50.
3. Foucault, *The Archaeology of Knowledge*, p. 82.

series of letters, A, Z, E, R, T, listed in a typewriting manual, is the statement of the alphabetical order adopted by French typewriters'.[4] From a less discourse-analytic and more media-archaeological perspective, any technological articulation is always already *parole*, not simply an ideal *langue*. Every symbol is materially glued to hardware as its condition of possibility to come into existence and to become operative at all. In that sense its technicity is irreducible and even redefines the relation between nature and culture. The symbolic order is never purely intransitive, but always already transitively entangled with the physical world. 'This pile of printer's characters, which I can hold in my hand, or the letters marked on the keyboard of a typewriter are not statements: at most they are tools with which one can write statements.'[5] What for Foucault is merely the instrumental medium gets a different momentum in transitive media philosophy, where every act of writing is consciously perceived in terms of Markov chains and information entropy. The non-alphabetical order of letters on the keyboard of a typewriter has been the result of a ratio (*lógos*) between mechanic limitations (respectively frictions) and stochastic letter distribution probabilities for the sake of speed-writing. Foucault's rather text-oriented discourse analysis tends to overlook this *machine statement*.

Friedrich Nietzsche was among the first philosophers to acquire the new medium of a mechanical typewriter. First of all, this had the simple function of coping with his near-sightedness via a mechanical writing aid. In a true moment of transitive media insight, though, one of the typescripts he produced epistemically (self-)reflects: 'Our writing tools are also working on our thoughts.'[6] Since his machine (specifically, the Malling-Hansen writing ball) did not allow him to recognize the imprinted letter at the moment of inscription, this actually enhanced *écriture automatique* and a telegraphic style. 'After abandoning his malfunctioning machine, Nietzsche elevated the typewriter itself to the "status of a philosophy," suggesting in *On the Genealogy of Morals* that humanity has shifted away from its inborn faculties [...] in favor of a memory machine'.[7] If such a core concern of the human–machine relationship is directly experienced at the very moment of machine writing

4. Foucault, *The Archaeology of Knowledge*, p. 86.
5. Foucault, *The Archaeology of Knowledge*, p. 85.
6. 'Unser Schreibzeug arbeitet mit an unseren Gedanken', as quoted in the translators' introduction (Geoffrey Winthrop-Young and Michael Wutz) to Kittler, *Gramophone, Film, Typewriter*, p. xxiv.
7. Winthrop-Young and Wutz, ibid.

itself (and not just written *about*), a philosophical moment of transitive media insight occurs.

Opening Ashby's 'Black Box'

A lot of interactive media-artistic installations actually dissimulate the technical tools and logical codes behind their apparently intuitive (mostly audiovisual) surface. This privileges aesthetic pleasure above their techno-analytic knowledge potential. The human observer is reduced to an affective participator, while a lot of artists leave a more transitive technical insight to the engineers and programmers they co-operate with. A notable exception is real-time video editing and live coding, which does not defer the creative momentum to post-production but involves it, like an anticipated future, already in the now-time of the event itself. Cultural metaphors (in the semiotic sense) are usually hiding the actual operations of the symbolical machine in its technical sense, just as if the visible pixels had no physicality. 'Mind the gap between intransitive and transitive media-aesthetic awareness', media archaeology responds. But even if the circuitry behind an LED monitor in a media artwork becomes transparent to the observer (such as in Branimir Štivić's ongoing installation project *Conveyor*[8]), its actual signal processing still remains occult for human perception. It takes non-human observers (e.g. an oscilloscope that displays the spectrogram of the signal event) to supplement such a deficiency.

Transitive media philosophy insists on opening the black box of actual technologies not only in critical (that is, textually distanced) analysis but by becoming aware of the technological momentum in real-time reflection. It operates with(in) the micro-infrastructure of devices that are otherwise experienced, by human users, as technically intransitive interfaces. Cybernetic experimenter and epistemologist William Ross Ashby once physically implemented a black box (known as the Ashby Box, completed in 1954) that was not meant to be opened at all. The relation between input and output seems unpredictable here. Yet Fischer and Cretu initiated a media-archaeological remake of the Ashby Box: 'Without insights into its inner workings and without performing a detailed analysis of its behaviour, we speculated how a generally comparable equivalent of the original device might be implemented using components that were common at the time of its implementation and featured in other works of its designer',[9] such as the uniselector (known from contemporary electromechanical telephone network switching).

8. See https://branimirstivic.com/traka.html
9. Fischer and Cretu, 'The Ashby Box', p. 182.

Figure 3.1: Uniselector mechanism for automatic telephone exchange.

The Ashby Box embodies a non-trivial machine (NTM) with its inaccessible circuitry,[10] 'for which we do not necessarily have any information of the structure by which this operation is performed'.[11] Its input switches perform two functions: 'Besides delivering the electrical on/off inputs, they also cause the Ashby Box to transition from its present state to the next every time they are toggled.'[12] The combinatorial space that unfolds between possible inputs and possible internal states results in multiple numbers of input–output relationships rendering the machine 'effectively unpredictable from an observer's perspective'.[13] But a hypothetical modelling by a logical truth table of possible transfer functions of the Ashby Box indeed corresponds with the wiring of a column of fixed contacts on an eight-bank uniselector.[14] Transitive media philosophy is thereby incited to muse about the mechanization of mathematics that is the principle of digital computing. Does such an

10. See von Foerster, *Understanding Understanding*, pp. 309–13.
11. Wiener, *Cybernetics*, xi.
12. Fischer and Cretu, 'The Ashby Box', p. 184. For a video demonstration of the Ashby Box (as part of a lecture by Thomas Fischer) see https://youtu.be/ETtJfr6fZvQ?t=2843 [accessed 10 October 2025].
13. Fischer and Cretu, 'The Ashby Box', p. 168.
14. See Fischer and Cretu, 'The Ashby Box', p. 183 (Fig. 3).

experience of an otherwise opaque mechanism come to its limits with artificial neural nets in current machine learning?

Soldering Electronics

On 16 August 2024, on the workbench of the Media Archaeological Fundus in the Department of Media Science at Humboldt University of Berlin, a mechanically programmable toy car, the *Kybernet* from the former GDR industry,[15] remained silent even after the operating battery was replaced by a fresh one. Only with systematic error analysis (voltage measuring) was it revealed that even the replacement battery had unexpectedly been exhausted in its storage latency over time. Any analysis of technical hardware or the mechanic notion of an actual machine is too narrow unless it also involves its framing by energy storage and signal transduction. It was not the internal system of the *Kybernet* but its external power supply that failed, directing attention to the common neglect of peripheral devices such as antennas for radio reception, or cables for interfacing media with each other. The apparently marginal device deserves a thorough media-philosophical reconsideration: The proper machinery, as a technical being, is already entangled with its periphery to become an operative medium, just like a frame is not external but belongs to the painting as a material operator distinguishing the figure from its ground.

In the *Kybernet* case, though, even the implementation of a third, active battery did not result in a machine action. Investigative techno-analysis finally revealed the error: An electrical wire connecting the energy source to the motor had been disrupted and required renewal by soldering. Soldering should not simply be regarded as a technical practice of repair for electronic devices but deserves to be consciously experienced as an active switch from abstract media analysis to transitive media synthesis. In media-epistemical real-time experience, soldering is revealed as the concrete encounter of techno-logical reasoning (the diagram) with matter and energy, from which a symbolic machine (the actual circuitry) arises. A whole world of media-philosophical speculations evolves from the creative awareness of such a transitive moment.

The most transitive kind of *writing the medium* is hardwiring its electronics as directly as possible. Its computational equivalent, for digital media, is programming close to the machine, such as in the time-critical Assembly code. Not only to *do* but even to *consciously reflect* this in real-time with

15. For further details see Salvaggio, 'Toy Cars and Cybernetic Communism'.

Figure 3.2: Inside and outside of the *Kybernet* programmable car toy.

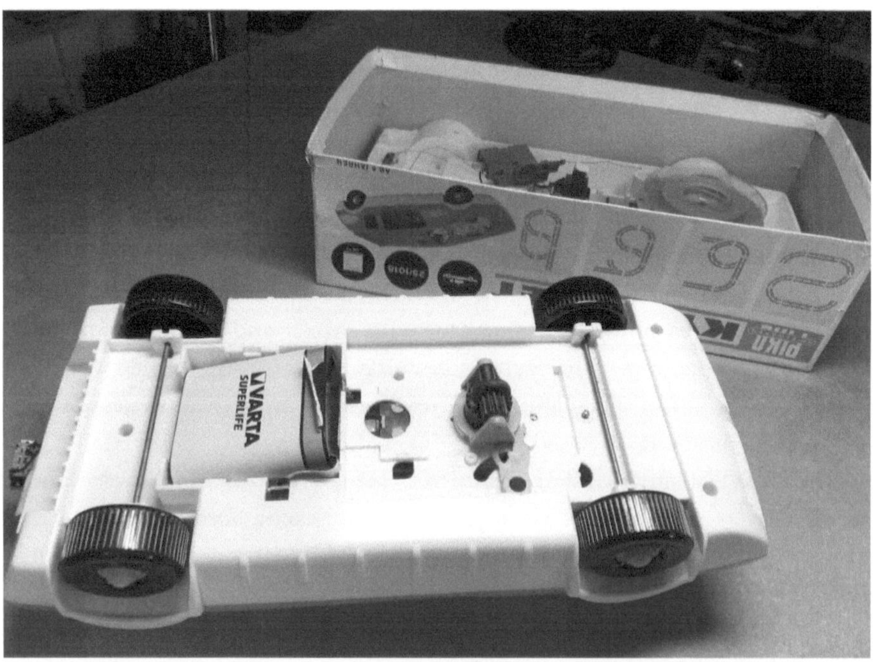

Figure 3.3: Battery supply of the *Kybernet* programmable car toy.

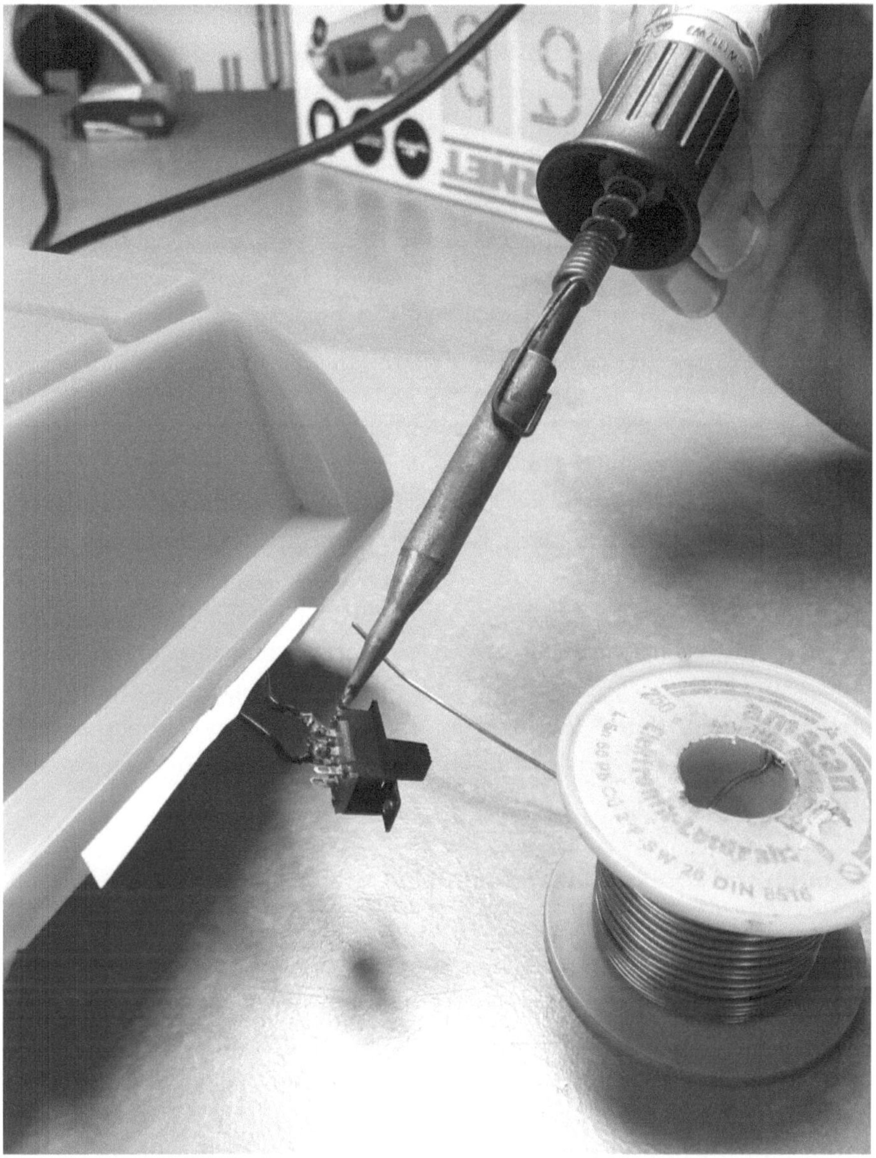

Figure 3.4: Soldering moment at the energy switch of the *Kybernet*, on the workbench of the Media Archaeological Fundus.

media-epistemic awareness beyond the immediate engineering is the perfect mode of applied media philosophy (in analogy to cognitive awareness during aesthetic production in the arts). Even for a dilettante in engineering, the act of soldering an electronic circuit is a most intimate media contact of logical reason with the technical real. The symbolically notated diagram becomes electro-materially existential here. An objectification of this transitive moment is the soldering iron itself. During such an encounter, there is a moment when it is uncertain whether there is still operative logic at work or already brute matter in its own physical (entropic) activity. With the smell of solder the logific(a)tion of matter becomes a sensuous experience.[16] The solid machine order here literally melts into air. Joining two metal surfaces together using a filler metal occasionally results in a so-called cold solder joint that 'may not conduct at all, or may conduct only intermittently'.[17] Transitive implementation of logical gates becomes resistant here, a kind of *transistor* momentum in its material sense.

A simple study of the circuitry design of an electric toy like the *Kybernet* can only conceptually but not concretely anticipate such kinds of material error that only arise in the moment of its mechanical realization. Digital computation is familiar with the impossibility of algorithmically calculating, in advance, its own completeness to perform its task with any given input data in finite time.[18] In 'deep' machine learning, for instance, the k-means-clustering algorithm configurates data points into groups but may do so occasionally in nonsense ways (from the human point of view).[19] What can be experienced by t(h)inkering with the concrete software code of such an algorithm is qualified as transitive media philosophy once it switches from intuitive playful attention into conscious media-epistemic insight.[20]

Through the Looking Glass: Cognizing the Medium Message Reflexively

It happened with the cultural technique of writing before it became electronic technology. In the manuscript epoch, and even with its acceleration by the mechanical typewriter, the temporal gap between flashes of thought as 'stream of consciousness' and their actual materialization in alphabetic letters, their hand-writing and final postal transmission still invited for moments of musing. This reflexive delay has collapsed almost to zero in times of telegraphy and instant messaging via mobile smartphones. In the 'digital natives' generation, the act of writing has nowadays advanced to the speed of affect.

---

16. For a more detailed description see Schulze, 'Löten', pp. 176–99.
17. https://en.wikipedia.org/wiki/Soldering [accessed 2 February 2024].
18. On such incompleteness see Fazi, *Contingent Computation*.
19. Noack and Sanner, *Künstliche Intelligenz verstehen*, p. 191 (subchapter 9.4.1 'Unsinnige Gruppierungen').
20. For an exemplary study see Berry, 'Tracing "Toxicity" Through Code'.

Transitive media philosophy aims at a nevertheless *conscious* epistemic awareness of that moment in real-time. It requires a media-epistemically biased mind, though, in order to perform such an original metareflection as coextensive with the experienced medium event. In a medium sense of such a 'reflection', one cannot only see something or be seen, but rather see oneself *in* and *by* the mirror in immediacy. Technology makes a significant difference here as well, with regard to McLuhan's focus on the actual 'message' of the technical medium rather than on its semantic, musical or iconic content. While an everyday smartphone is occasionally used for cosmetic facial self-control, the transitive media philosopher is aware of, and consciously reflects, such a scene as an instantiation of the subliminal medium message. Differing from the analog mirror image, the 'selfie' is already determined by its imagined destination as internet 'post'—that is, by its technical condition of possibility for transfer in the electromagnetic channel. While a self-image in the catoptric mirror is physically reflexive, i.e. transitive, the dioptric (transparent, diffractive) electronic video screen renders an always already technically processed, i.e. intransitive, portrait. Optical refraction becomes epistemically autopoietic once it is reflected in the transitive media-philosophical sense.

A similarly 'transitive' scene that is known from the cultural techno-imaginary: A film spectator may drift into the screen, 'much like the legendary Chinese painter Wu Daozi who, longing for the peace of the landscape he had created, moved into it, walked toward the faraway mountains suggested by his brush strokes and disappeared in them never to be seen again'.[21] System theory elevates such a moment to epistemic dignity in terms of second-order cybernetics.[22] Media artist Dan Graham created double-mirror rooms to make the observer aware of their own observation. While an optical reflection usually remains a rather intuitive interactive 'live' experience for visitors, transitive media philosophy aims at turning it into conscious media-induced insight in real-time. Graham, in his installation *Present—Continuous—Past(s)* from 1974, extended his mirror installations via a magnetic tape recording loop and video monitors for time-delayed observation. The media-epistemically trained mind immediately turns such an aesthetic experience into a conscious remembrance of, for instance, Heinrich Hertz's experimental mirror reflections of radio waves. Next to such scientific considerations, deep insights into the processual media-nature of passing 'time' itself arise.

21. Kracauer, *Theory of Film*, p. 165.
22. As has been elaborated in von Foerster, *Understanding Understanding*.

Machine Phenomenology: Experiencing Rembrandt TV

To what degree are technological objects 'real things whose core reality is withdrawn from access'?[23] As culturally informed artefacts of physical matter and energy, they exist at the intersection of mindful intelligibility and an idiosyncratic knowledge of its own media-specific kind. This ambivalence is irreducible. Technically in/formed matter, 'far from being a passive stuff awaiting our formation or instructions instead modifies our designs in all sorts of unexpected ways. [...] The inventor of the clock did not intend for it to striate every aspect of life.'[24] Technical objects are accessible as 'epistemic things',[25] either by humans or even below the human cognitive threshold by artificial neural nets themselves, once they have been explicitly trained to develop an epistemic sense for tec-knowledge (an interest that has so far escaped the big data AI companies). In a concrete rather than 'speculative' realism, technical *lógos* is doubly bound to its materially in/formable and to its mathematically computable existence as operative diagram. Technical beings, once perceived in transitive rather than abstract philosophy, become the media scenes of a material experimentation of reason that can be experienced, tested and recognized as media-ontological agencies. The actual encounters between techno-philosophical knowledge and its concrete physical engineering objects result in a micro-laboratory for experiencing the technical world from its own perspective. To a certain degree in analogy with object-oriented ontology but much more techno-logically specific, this approach does not privilege human reason over the existence of material knowledge that is not considered passive once objects have been techno-reasonably informed. Their latent tec-knowledge options are not limited to their pragmatic human usage but invite transient philosophical awareness to become epistemically explicit.

In actual technologies, the worlds of both human and machine reasoning are rather equi-primordially entangled. To take this one step further, *process*-oriented media ontology maintains that technical objects tend to develop an independent knowledge of their own, even if they have been shaped by human invention and engineering as reifications of the cultural mind. This reminds of the *noumena* in Kant's philosophy, but less ideally: A technological device becomes concretely accessible and recognizable as thought object once its electronic existence is transitively reflected as such.

23. An expression borrowed from Morton, *Hyperobjects*, p. 15.
24. Bryant, Onto-Cartography, p. 22.
25. In the sense of Rheinberger, *Towards a History of Epistemic Things*.

One case study is provided by the veritable machine revelation that results from the reactivation and white-boxing of a first-generation GDR televisor from 1954 (trademark 'Rembrandt'[26]). Such a technical re-enactment allows the difference between the description of a communication medium in a manual and its actual collateral signal behaviour to be experienced, and makes the philosophical debate between idealism and materialism become concrete in a technological instantiation. For example, there is an acoustic tone arising from the high-frequency scan lines that unfolds from the vibrating grid wire within the oscillatory cathode ray tube.[27] At first this may appear involuntary or even as collateral damage. To the media-epistemically tuned mind, though, the electronic 'image', beyond its manifest iconic content, all of a sudden reveals its implicit sonicity.[28] This triggers a whole chain of thoughts about the privileged alliance between the regime of electromagnetic waves and the temporal essence of sounding signals. Watching classical cathode ray TV indeed leads to transitive media experiences: With the picture tube, one of the basic electronic elements that are usually hidden in the case becomes the actual interface to human perception. The conventional dichotomy between user-oriented media-phenomenal surface and its electronic truth is subverted in a literally enlightening technical mode.

As electronic screen, the cathode ray tube breaks through from the inside of the functional internal electronic elements to the surface. In contrast to cinema projection, the electronic cathode ray that emanates from the deep dioptric television screen 'writes' an electronic image that is both sequential and discontinuous with its scan jumps. The electronic 'scanning finger' results in a direct *massage* of the retina in the human eye,[29] while the digital LED screen image nowadays is rather intransitive to the computational data processing that occurs within the internal microchip circuitry, unless pixel-wise disruptions ('artefacts') occur from erroneous signal transmission. The transitive media philosopher, while watching analog TV, is not only precisely informed and critically aware of such technical differences but instantly reflects them against the metaphysical quest for immateriality in the (nevertheless substantial) optical appearances that have dominated visual history so far. Media-active electronic *imaging* is not simply a technical instantiation or escalation of a

26. FE852E, VEB Sachsenwerk Radeberg. For the Rembrandt TV in its original shape, with its electronics still hidden within the case, see https://wikis.hu-berlin.de/maf/TV_Rembrandt#/media/Datei:IMG_166-1.png [accessed 10 October 2025].

27. Jakubaschk and Scholz, *Fernsehempfänger selbstgebaut*, p. 8.

28. This becomes explicit in Viola, 'The Sound of One Line Scanning', pp. 39–54.

29. McLuhan, *Understanding Media*, p. 313.

Figure 3.5: Rembrandt television set (GDR 1954) 'in being' with the media-epistemically participant observer.

transcendent image culture, but epistemically challenges its anthropocentric image ontology. The electronic image is an *image act*,[30] in its technical sense, through knowing signal formations and data formats, but no 'image' itself. In the consciously perceived wonders of the electronic screen, an intellectual insight into technology occurs as technical 'illumination' literally.

Philosophizing with the Magnetic Tape Recorder

'Carpentry' has been proposed by Ian Bogost as an experimental method of *object-* and *process-oriented* media-ontological analysis. Such a materially transitive media investigation is in accordance with 'philosophizing with the magnetic tape recorder'.[31] Such an approach does not simply focus on a technical description of its time-manipulative medium specificity (against the recording-only phonograph) but perceives it as an electronic device to reveal alternative concepts of sound as an 'epistemic thing'[32] that even bypasses explicit acoustics and human hearing.[33] Inversely, the transitivity of a magnetophone's

30. A concept developed Bredekamp, *Image Acts*.
31. Haberer, 'Phonographendämmerung?', p. 275.
32. A notion coined by Rheinberger, *Towards a History of Epistemic Things*.
33. As expressed by Meyer-Eppler, 'Elektronische Kompositionstechnik', p. 7.

Figure 3.6: Ripped driving belt within a magnetophone.

inner mechanism is revealed in moments of malfunction, such as a ripped driving belt that links the electric motor to the actual tape reels.

If such a friction is addressed not only in terms of functional repair but with media-epistemically tuned attention, the technical situation triggers philosophical speculations about the relation between electronics and mechanics in technical devices. There is an idiosyncratic tempor(e)ality arising from the various materialities within such an electronic apparatus that deconstructs the techno-logical idealism of its more abstract circuit design. Media-archaeologically such a disruption raises the question of how a worn-out rubber ribbon relates to the magnetic storage medium as such: Is such material entropy peripheral or integral to the definition of technology? Any logocentrism in media-cultural philosophy thereby becomes deconstructed.

Knowing Sound and Music Techno-Logically

With sounding media, technical refinement and aesthetic perception coincide. *Transitive* media philosophy turns such an experience into epistemic awareness (*aisthesis*). As *transitive* machine, the analog electroacoustic synthesizer is directly operated like its technical twin, the analog computer. The *analog* (as a technical term from electronics) refers to the actual circuitry here, while the *analogue* is the cultural concept of music as applied to such machines. 'Get to know the machine,' declares sound artist Danielle 'Danz' Johnson on her usage of modular Moog synthesizers. Their inherent sonicity longs to be revealed as a function of their considerable technical qualities. The careful study of the manual of such a sound-synthesizing machine (its techno-logical information) definitely grants 'an added benefit of knowing the ins and out of the machine',[34] while at the same time no sound as such will arise from the manual. The actual machine experience turns into conscious 'thinkering' (instead of simply intuitive trial and error) once the user is not only hunting for the implicit musical ratios (*lógoi*) but also for the media-epistemic *lógos* that resides within the material circuitry of the machine.

Media-scientific performance lectures, such as the knowledge-oriented demonstration of an electroacoustic synthesizer, may thereby express a critically reflected 'dialogue with the machine' (in Rekveld's sense, as discussed above). The dialogical effort (even if still asymmetric) is immediately rewarded by actual sounds and noises. In most cases of media-artistic research the actual event does not already derive from an external artistic idea but is *induced* (even in its electrical-engineering sense) from within the machine. Transitive media-philosophical insight occurs from such a direct technical contact; the haptic and the epistemic dimension fold into each other. Its reasoning is attentive for such moments that can be immediately identified as articulations of techno*lógos*. To become conscious of such tec-knowledge, though, requires an epistemically *tuned* mindset.

34. Johnson, 'Alive and Analog', p. 88.

Haptic Experience of Sonic Media as Technology

As a transitive, immediate and epistemically contagious enactment, medium-philosophical reflection is not immaterial or a pure idea that takes place in a Platonic sphere. Any abstract theory is untrue against technological reality. Concerning approximations to inner-technical tendencies and singularities, an epistemologically path-breaking philosophy is demonstrated by Aristotle's media-specific differentiation of the human senses: 'For each sense there is a medium (metaxy), which performs a decisive function: for sight, the medium is the diaphanous that, illuminated by color, acts upon the eyes',[35] such as the Maria glass crystal. This physical approach to light has been replaced by the detection of spectral electromagnetic waves that propagate themselves immediately. For hearing, though, it is still 'the air that, moved by a sounding body, strikes the ear. What distinguishes touch from the other senses is that we perceive the tangible not "because the medium exercises an action upon us, but together (*ama*) with the medium". This medium, which is not external to us but within us, is the flesh (*sarx*).'[36] The technological sphere now replaces such an embodiment with medium cognition: In transitive media-philosophical coupling to the technical device, the human mind is not simply affected but becomes self-reflexive through its reified other. Haptic contact with a technical artefact as objectification of the mind is therefore recognized as a kind of contact with materially implemented *lógos*. In the sense of an immediate transition (not theoretical 'inner sight' only), technological insight is generated from a concrete modality that cognitively immerses into the artefact.

Circuit-sensitive acoustic research becomes transitive with sensors such as the pressure points for voltage control in electroacoustic synthesizers, which absorb the interacting human into the cybernetic loop. The humidity of the human finger (or its extension by a metal stylus) itself modulates the resulting signal, as has been applied by Finnish electronic music pioneer Erkki Kurenniemi for the DIMI-A (in his Digital Music Instrument series).[37] By touching copper surfaces the player is no longer just the external director of the acoustic event but becomes part of the electronic circuit itself. Any such direct technical experience, also found for instance in the electroacoustic Trautonium, reveals the device as an 'object of transition where circuit and

---

35. As paraphrased by Agamben, 'A Philosophy of Contact'.
36. Agamben, ibid.
37. See the entry '"DIMI" & Helsinki Electronic Music Studio, Erkki Kurenniemi. Finland, 1961', *120 Years of Electronic Music*, https://120years.net/dimi-helsinki-electronic-music-studioerkki-kurenniemifinland1961-2 [accessed 24 February 2024].

matter are still symbiotically entangled'.[38] This is not simply a symbiosis in the sense of human-to-machine coupling but also occurs categorically: as encounter between the mate*real* and the symbolic. Both categories exist as rigid dichotomy only for the sake of differential analysis. In technological reality, they are always already intertwined but not to the degree of complete indifference. There is always a material and energetic reference for reason to unfold techno-logically.

Sounding Media Epistemics (Monochord, Electroacoustic Synthesizer)

Any transitive epistemology of technical devices media-archaeologically roots in processual experience. Such a dynamic experience is implicitly sonic in the sense of (almost) immediate resonance, while its visual perspective remains rather distanced.[39] Such a kind of techno-philosophical enquiry started with acoustic research that revealed the implicit mathematicity of the sonic medium. Mathematical theory is no external *in*transitive symbolic notation here but is derived from the technical operation itself.

Paradigmatic for such an artefactual self-articulation has been the ancient Greek monochord: an instrument which has not been applied for musical amusement but used as a laboratory device. The Pythagorean 'acoustemic' sonification of integer number ratios allowed for an immediate, transitive reflection of a philosophically supposed mathematically structured world harmony. Passing a blacksmith workshop, Pythagoras (as purported in Boethius' *De institutione musica*) recognized the harmonics sounding from the interplay of several hammers. Pythagoras did not subject the sonic instrument to a presupposed musical theory but listened to the proportional scaling of the monochord with 'acoustemic' ears. He thereby inductively derived that aesthetics was rational and that human psycho*lógos* is technically enabled to recognize the *kosmos* of proportional numbers.[40] Listening thereby became a sonic form of philosophical speculation. Such a techno-epistemic understanding is no longer dependent on a divine intuition or illumination but becomes a media-specific 'tuning' (to remain within the sonic metonymy). Perhaps less analytically but still media-specific, the special ways of playing

38. 'Das Trautonium ist ein Objekt der Transition—hier hängen Schaltung und Material noch symbiotisch zusammen.' Dörfling, *Der Schwingkreis*, p. 250, and Fig. 43 with a snapshot of a testing circuit.
39. As described for the sensation of atomic vibrations by Soentgen, 'Atome Sehen, Atome Hören', pp. 97–113.
40. Rieger, *Helmholtz Musicus*, p. 54.

Figure 3.7: Educational replica of the Pythagorean monochord (engineered by Ingold Haedicke).

the Chinese ch'in lute (its 'ideology'[41]) have also been a means of conscious immersion into the unity of cosm and instrument in the enactive moment. Such an *instant* philosophical awareness, in its material structure, formal

41. Van Gulik, *The Lore of the Chinese Lute*. Thanks to Dr Jingwei Wu (School of Journalism and Communication, Tsinghua University, Beijing) for this reference.

shape and skilled playing technique, is supposed to embody the supreme natural lessons of the Dao. 'Although the tones of the Lute may be featured, when listening to them who shall be able to fathom their significance?', Van Gulik quotes, on his book cover, from a poem by the Sung epoch scholar Ou-yang Hsiu. It therefore requires a 'cosmotechnically' tuned ear to listen to the ch'in's medium message.[42] Does an electronic music synthesizer that drastically differs from the body-related technique of playing the lute still embody the Dao or make this relation intransitive? It is a different kind of *lógos* that emanates from technological electroacoustic devices when compared to cultural techniques of playing musical instruments. The arbitrarily constructed monochord already has already instantiated a hardwired musical theory. But when the resonant instrument body is fully replaced by the electric circuit, a different kind of inner-technical knowledge may be revealed by transitive media-epistemic listening.

Transitive Media Philosophy and/or Lucier's Medium-Reflexive Aesthetics

Gordon Pask's electromechanical Musicolour Machine has been an example of knowledge-oriented media research art. This classic implementation of an autopoietic system allowed for the musician to interact with the mechanism in a way that synaesthetically connected the audio input to the visual output.[43] Through hardwired machine learning, the trainable Musicolour dynamics adapt to its human player, and via its flexibility draws the performer 'into novel epistemic and time-critical relationships imparting an increased media-epistemological sensitivity, as it were. […] This process […] operates within a paradigm of critical intimacy rather than the critical distance usually associated with art appreciation.'[44] That of course means immersive media aesthetics instead of cold media-theoretical observation.

Alvin Lucier's 'oscillating and processual-reflexive experimental aesthetics' also realizes an 'artistic philosophy'[45] that reveals otherwise inaudible, implicit sonicity[46] in vibrational events by technical sonification. Acoustemic hearing thereby turns into literally *understanding media*, even if such a method of

42. In the sense of Hui, *The Question Concerning Technology in China.*
43. The author owes this special focus of attention to Ronit Ghosh's article 'Critiquing Embodiment: The Technological Roots of "Biomusic"', forthcoming in the journal of music and technology *Organised Sound.*
44. Pask, 'A Comment, a Case History and a Plan', p. 80.
45. Rietbrock, *Alvin Lucier's Reflexive Experimental Aesthetics*, p. 224.
46. On this concept see Ernst, *Sonic Time Machines.*

cognitive *aisthesis* is still anthropocentric in its focus on the human ear. While a site-specific electroacoustic installation such as *TWINS / Music on a Long Thin Wire* (moderated by Lucier) conveys its inherent sonic knowledge as a rather sublime aesthetic message to the audience,[47] transitive media attention is immediately aware of, and epistemically ponders, the electromagnetic law of induction that conditions the sonic event but usually remains distant (even hidden) from human observation in acoustic space. It is exactly the shift of attention from the secondary acoustic effect to the primary electronic agency that differentiates transitive media-epistemic attention from pure listening. In Lucier's installation, the media-ontologically delicate transduction from mechanical to electric oscillations derives from the interaction of alternating current, the long dual metal wires, the responding horseshoe magnets and the contact pick-up microphones that shift authorship towards the technical apparatus itself—even if the portrait of the human artist is still shining through. Lucier's resonant double-wire installation acoustically results in subtle timbres. Beyond such aesthetic pleasure, a more transitive, epistemologically attuned awareness of that medium event, though, immediately speculates on the non-linear and unpredictable ('chaotic') behaviour that may result from such a system configuration as tec-knowable event.

While Lucier's original installation (conceived in 1977) remains an electronic extension of the Pythagorean monochord, its twin duplication in Berlin (still triggered by the original oscillator) resulted in a musical escalation: It embedded this update in a millennium-old acoustemology 'as a way of knowing',[48] which even includes the duality of the Siren voices in Homer's *Odyssey*.[49] When is it justified to term electromagnetically induced resonance and cybernetic feedback 'music'? How does this cultural symbolism relate to its technical ground, implying an implicit tec-knowledge? Such media-operational flashes of enquiry do not emanate from theoretical deduction but are induced by the concrete physicality of a spatiotemporal scene. Once such a moment is not aesthetically reduced to passive immersion but recognized in cognitive epistemic awareness, it becomes explicit insight on the spot. This experience may even be preserved and expressed, in an extended presence, in the delayed textual writing scene.

---

47. Original installation at Parochial Church Berlin 1999; reinstallation 11–22 September 2024, as part of the singuhr projekte festival *Extended Spaces—Resonant Bodies: Alvin Lucier*.
48. Feld, 'Acoustemology', p. 12.
49. See Frommolt and Carlé, 'The Song of the Sirens', pp. 18–33.

In Lucier's seminal media performance *I Am Sitting in a Room* from 1969, the technical ground (the monophonic electromagnetic tape recorder) hides behind the figurative 'alien phenomenology' of its auditive experience. Complementary to such acoustemic second-order formations of operative devices in spatiotemporal installations, transitive media philosophy shares the instant momentum but directs attention to its first-order inner-technical condition too: in/formatized matter as concretization of the Kantean *a priori* and the Foucauldean *l'archive*. To modify a line from William Carlos Williams' poem 'A Sort of a Song' ('no ideas / but in things'[50]), there is media evidence that is not logocentrically imposed upon the technical event as its theory but emerges from its actual signal processing *in being*. When the musical scores are identified as already electronically inhe(a)rent 'in the circuitry' itself,[51] the symbolic suddenly becomes transitive *to* (and *as*) the technically informed (mate-)real. Media-aesthetic experience, in such a transient philosophical awareness, can thereby be made coextensive with its reflexive tec-knowledge, even if this remains an asymmetrical relation. Technical sonicity is a privileged field of such an experi*mentalization*.

Transitive Radio Acoustemics

In the double sense of ancient Greek *akouein* for close listening, a transitive media epistemics of prosthetic hearing aids like the cochlear implant pays attention to machine listening itself. Such an analysis of this neurotechnology not only provides for media-archaeological insights *into* and *about* (literal media *theoría*) but acoustemically listens to 'Insights *from* the Cochlea Implant'.[52] As a rather unintentional side effect this neurotechnological device occasionally sonifies environmental electromagnetic noise that is otherwise inaccessible to regular human hearing.[53] The familiar acoustic environment becomes redefined by the 'acoustic space',[54] radio waves that not only extend but fundamentally transform the sphere of cultural communication. Such an

---

50. See Hauke Harder and Viola Rusche's film *No Ideas but in Things: The Composer Alvin Lucier* from 2012: http://www.alvin-lucier-film.com/moaltw.html [accessed 22 September 2024].
51. Lucier, 'Origins of a Form', p. 6.
52. David Friedrich, 'Neurotechnologies and Their Hardwired Politics: Insights *from* the Cochlea Implant' (publication forthcoming in the journal *Techné: Research in Philosophy and Technology*). Friedrich has designed an algorithmic being that techno-logically implements hearing through a cochlear implant: *Naughty Box*; see https://youtu.be/qJbU87jJjpk [accessed 10 October 2025].
53. Ochsner and Stock, 'Das Hören des Cochlea-Implantats', p. 415.
54. On in the media-epistemic sense of this term beyond audible experience, see Carpenter and McLuhan, 'Acoustic Space', pp. 67–69.

Figure 3.8: Type-identical specimen of Heidegger's radio from the user perspective.

immediate awareness of a different media knowledge requires a direct coupling with the technical event: less on its phenomenal surface level (the user interface) and more with the acoustemic implications that arise from bare inner-technical introspection into the essence of radio technologies. Heidegger's critique of intransitively tuning into the world via radio once derived from his user experience of turning the radio knob.

This user-centric approach is opposed to a more direct media-archaeological experience of radio reception once the presence of the electromagnetic frequency oscillators (first thermionic tubes, then transistor circuits) is taken into account of defining communication. In radio electronics, the 'receiver' is not people in the sense of communication studies but rather the resonating circuit within the apparatus, long before the frequencies are finally transduced into acoustic signals that can be identified by human ears.[55] Communication is thereby identified as primary technical rather than social.

55. For that reason the Media Archaeological Fundus (MAF) in the Department of Media Science at Humboldt University in Berlin actually contains two type-identical versions of the radio that Heidegger once housed in his legendary hut: one in its familiar user-oriented design, and another with its electronics laid bare.

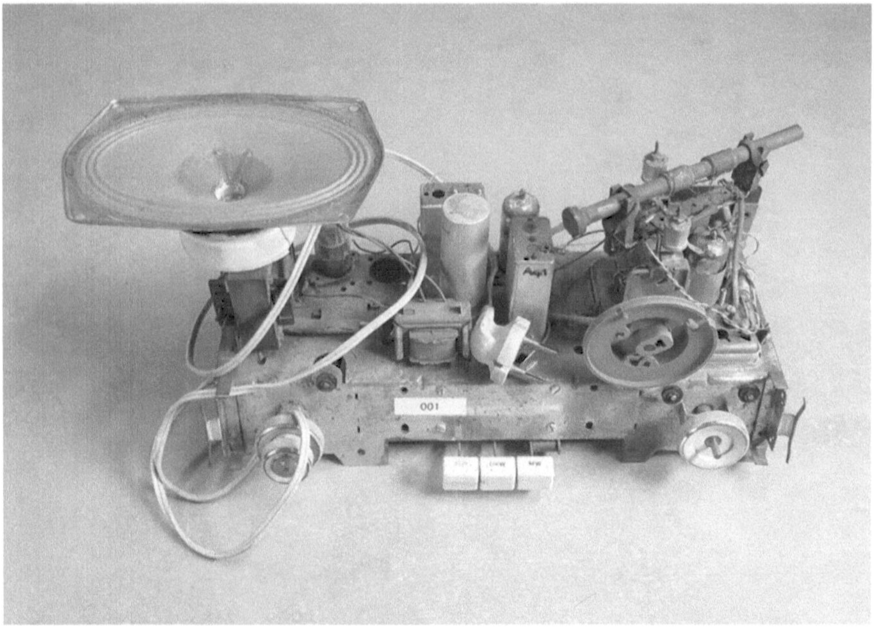

Figure 3.9: Heidegger's radio with its inner electronics 'revealed'.

The essence of radio has been epistemic long before it became known as a primary agency of mass media communication. The phenomenon of electromagnetic induction itself is charged with all kinds of techno-epistemic knowledge, from the physical to the conceptual aspects: It was discovered by Oersted spontaneously, then systematically experimented with by Faraday and subsequently mathematically calculated by Maxwell. Finally it was transformed into electromagnetic wave oscillations by Hertz as the condition of possibility for what subsequently became the mass medium of 'radio' transmission. Transitive media philosophy, whenever listening to 'radio', rather muses about its multiple techno-epistemic implications than its verbal or musical content. Beyond the otherwise affective, rather non-conscious moment, it cognitively experiences radio's multiple layers of tec-knowledge at once. When listening to analogue AM short- or medium-wave radio, the reception is not clearly focused on the verbal, musical or Morse-coded content alone but insistently noisy. A media-archaeologically tuned understanding that resonates with 'the ears of the machine' explicitly listens to such distortions to identify the technical medium message behind its semantic content. The ionospheric physics of electromagnetic radio transmission itself always remains co-present in the radio aesthetics of short-wave reception. This is a direct, in fact transient radio experience,[56] where the technical essence of communication is not digitally dissimulated in favour of a purified signal fidelity.

Such a transitive awareness of radio as structural medium besides its semantic content became most concrete with the claim for a technical feedback channel (against monopolist broadcasting) in Bertolt Brecht's 'radio theory' (1932). And in his inaugural speech of 22 August 1930 to the 7th German broadcasting media exhibition (Deutsche Funkausstellung und Phonoschau) in Berlin, Albert Einstein media-philosophically did not simply talk *about* the (still-) new wireless medium in the presence of a live audience. He instead made the tele-communicative public aware of its transitive radio immersion, with the radio being both subject and agency of his speech propagation. Einstein addressed the essential difference that an electromagnetic wave-based communication makes as compared to a classical audience in corporeal 'social' presence. Radio as technical event, in its correlation of technical medium message and semantic media content in Einstein's speech, became immediately reflected in the speaker's ironical salutation in its spatio-temporal hybridity: 'Dear Presentees and Absentees'.[57] This initial address captures the essence of radio: its phenomenology of the disembodied voice that is transduced into an electrical signal for transmission. *Understanding media* becomes technically concrete here with the accent on technical activity. Additionally, the transitive media-philosopher, when listening to such a document from the media archive, reflects upon the storage technology that made today's re-enactment of Einstein's ephemeral speech possible: phonographic recording on a solid medium. 'Live' spatial transmission and temporal replay become correlated here, resulting in media-epistemic flashes of critical questions concerning the conventional dichotomy between transfer and storage as media functions. Deferred transmission, as becomes technically concrete in material *delay lines*, embodies both qualities.[58]

---

56. Close to 'natural radio': Kahn, *Earth Sound Earth Signal.*

57. 'Verehrte An- und Abwesende': *Albert Einstein in the World Wide Web*, https://einstein-website.de/tondokument [accessed 15 September 2024].

58. On the engineering aspects see Sharpless, 'Mercury Delay Lines as a Memory Unit', pp. 103–09.

Chapter 4

Critical Outlook: Borderlines of Transitive Media Philosophy

Artificial Intelligence as a Challenge for Transitive Media Philosophy

A technological gap separates transitive hands-on design, analysis and immediate epistemic awareness of electronic circuitry from automated very-large-scale integrated circuit technology that withdraws from immediate human perception. In combination with complex algorithmic procedures, computing media thereby become techno-philosophically transcendental. They actually exist but escape immediate human cognitive experience and delegate their knowledge to distant media-theoretical criticism and sense-making instead. On a further organizational level, such an intransitivity escalates with the complexity of artificial neural nets for 'deep' machine learning to generate the effects of an artificial intelligence. The paradigm shift from classical computing to large language models within the Transformer architecture (resulting in applications like ChatGPT) occurs as an abrupt acceleration rather than a smooth evolution. Technical events become opaque idiosyncrasies. Their inner sequences seem transparent only to the awareness of techno*lógos* itself, or to automated procedures to make AI explainable. Can there still be a dialogue with this mechanism? In a world where everyday communication between humans becomes increasingly replaced by human communication with smart bots, human cognition itself becomes more than ever used *to* and formatted or conditioned *by* the language (techno*lógos*) of 'intelligent' software. Rather than simply criticizing this anthropological offence, transitive media philosophy is understood as a training of the 'user' to become aware of the epistemic beauty of this fascinating redefinition of the human–machine relationship.

Classical *technical* media, in their electronic circuitry, could still be transitively experienced as signal-transmitting and -storing devices in 'analog' days. It was their additional symbol-processing faculty (algorithmic computing) that turned them into more intransitive, truly *technological* devices. While the computer, conceived as symbolical machine, externally mapped code on matter, AI technologies nowadays seem capable of symbolization by themselves. Machine learning does not simply intransitively *simulate* but *emulates* mental brain processes, down to the subsymbolic signal transduction that occurs in the electro-chemical wetware of the human brain. Whatever *lógos* is generated and synthesized from the plasticity of artificial neural nets that have been trained by big data pattern recognition in 'deep' machine learning rather becomes a function of technically informed matter itself. Such media are not simply materializations of the cultural mind but a different kind of bit-critical reason. 'They communicate an Other'—in fact, a techno*lógos*—'that evades direct access, something that is neither only sign, nor only perception, nor only representation.'[1] In artificial neural nets, the machine-learning algorithm, depending on the training data, is able to modify itself by recurrent backpropagation. As has been expressed in the 'Manifesto' by the media-artistic creators of the AI-generated portrait *La Famille de Belamy* (2018), with big-data-trained art, the conceptual idea (design or score) no longer precedes its material in/formation, as with classical imperative computer programming. There is rather a reciprocal coevolution 'where the hands of the artist and the one of the machine are joined'.[2] For philosophizing not only *about* but actually *with* AI, the field of artificial-neural-net-based generative art invites a thorough understanding of how this technological being actually works, since such AI art 'is tangible: it offers some concrete results when experimenting'.[3]

At first sight, the explosive development of AI tools is bringing the traditional philosopher of technology into a more than ever intransitive relation with the inner-technical event. But just as in media semiotics even the digital image can still indexically relate to its external objects,[4] a second-order transitivity may arise as a shift of epistemic awareness from direct engineering to the techno-mathematical level that is the medium message of 'deep' machine-learning tools. Even on the user interface level, another kind of *transitive* relation to the computational event takes place. Instead of distant thinking about media as theory, the input practice of 'prompt engineering' becomes a literal dialogue with the computational mechanism itself. What has been experienced as an 'alien phenomenology' of electronics so far (Bogost), all of a sudden turns into an intellectual partnership—that is, into a co-reasoning of *artificial* intelligence.

1. From the present book series proposal by Bernd Herzogenrath and Tomáš Dvořák (draft).
2. Obvious, *Artificial Intelligence for Art*, p. 2.
3. Obvious, *Artificial Intelligence for Art*, p. 6.
4. As has been emphasized in Marks, 'How Electrons Remember'.

The price for such sophistication, though, is an ever more unbridgeable gap between the dialogical interface insight and the actual unexplainability of concrete decision-making on the techno-logical (both material and mathematical) level of artificial neural nets in combination with vast amounts of data. Its 'emergent' phenomena remain almost unpredictable in their idiosyncrasies.

Are AI tools generating a new computational analphabetism, just like the seminal Apple MacIntosh personal computer once did when introducing its 'Catholic' icons on the user interface against the 'Protestant' scriptural IBM command lines (in Umberto Eco's sense[5])? As expressed by a user's 'final thoughts', a tool like Websim AI allows the generation of complex web design through simple prompts 'even for those without technical backgrounds [...] to create and customize pages on the fly [...] allowing you to focus on creativity rather than coding'.[6] Transitivity in the relation between the human and the technical object becomes a dissimulation of its very mechanism. The address of the medium is its very message already: https://websim.ai. Can the (re-)emergent technology of machine learning, as it becomes ubiquitous computing in formats whose code and hardware is only partially disclosed by their enterprises, be epistemically reflected while it develops? Or does it require a temporal postponement, for the historian of technology, to describe it in media-theoretical distance?

Transitive media philosophy aims at insight towards such phenomena on the fly. As has been advocated in this book's argumentation so far, the transitive contemporary media-epistemic momentum occurs in the current epoch as long as its actual technological archive has not yet been historicized. Only this media-archaeological situation allows for an epistemic resonance with a new medium *in statu nascendi*. The analytic focus here is on its media-specific symptoms. In AI-generated video sequences, the asymmetries between the idiosyncratic techno-logic of computing and more traditional media of the moving image are increasingly smoothed. But it is especially with ephemeral errors and its 'glitches' or surrealistic distortions that the different mechanism of AIsthetics is occasionally revealed.[7] A time-critically transitive, instant media-epistemic awareness is required to catch and turn such momentary

5. See 'Umberto Eco on Macintosh vs DOS', an English translation from Umberto Eco's back-page column ('La bustina di Minerva') in the Italian newsweekly *Espresso* (30 September 1994): https://jowett.web.cern.ch/EcoMACDOS.htm [accessed 19 November 2024].

6. Hill, 'Creating a Website with Websim AI'.

7. This neographism has been borrowed from the symposium title *AIsthEthics (Artificial Intelligence, Aesthetics, Ethics in Art)*, University of Rijeka (Center of Innovative Media), 17 October 2024.

phenomena into an inspiration for media-philosophical enlightenment. The audio synthesis platform Demiurge, for instance, is driven by machine learning. A call for electronic compositions based on such material resulted in the Debris project that formalizes the creative recycling of digital audio by-products.[8] Once such AI contingencies are transitively *understood* (both heard and decoded) as indexical figures against an epistemic ground, an experimental working with such proto-aesthetic 'waste' positively turns into a media-philosophical toy for redefinitions of sound and music as such.

Can an increasingly intransitive technology still become accessible for a transitive media philosophy, as it becomes concrete with the analytic claim of explainable artificial intelligence (XAI)? The degree of mathematical abstraction in 'deep' machine learning clashes with the imperative of opening the black box in classical media archaeology and its related hardware tinkering. The haptic approach to technology becomes incommensurable with the complexity of mathematical models and their big training data. The multi-dimensionality of artificial neural nets is not simply electrotechnical but conceptual as well. The hardware *a priori* gets thereby redefined by techno-diagrammatic reasoning.[9] A direct transitive experience of a technical mechanism is frequently prevented by the 'protected mode' that is sealed hardware or archivally inaccessible records.[10] There can still be an indirect media-archaeological whitening of the black box by 'abductive' reasoning,[11] as a kind of techno-logical transitivity that makes such an opaque non-trivial mechanism symbolically transparent or at least 'porous'.[12] For AI-generated video outputs like those from Midjourney, the visual sequence is still triggered by text 'prompts'. With such a logocentric approach the word-to-image relation remains intransitive in linguistic terms.[13] In transitive media philosophy, though, the outputs of the technical device themselves are explicitly perceived, by its human opponent, as an expression of its inner implicit tec-knowledge. AI-generated texts, sounds and images are identified, by the

8. Trillo and Poliks, 'Debris', abstract.
9. On the notion of the 'techno-diagram', see Diego Gómez-Venegas' PhD thesis *Forecasting the Present: A Media-Archaeo-Genealogical Inquiry into Project Sybersyn* (Faculty of Cultural, Social and Educational Sciences of Humboldt University of Berlin, November 2023), https://edoc.hu-berlin.de/handle/18452/29888
10. In the sense of Kittler, 'Protected Mode', pp. 209–18.
11. For this term and heuristic method, see Peirce, 'Pragmatism and Abduction', pp. 112–31.
12. In the sense of Berry, 'Porosity and Computation'.
13. See Radford et al., 'Learning Transferable Visual Models from Natural Language Supervision'.

epistemically tuned observer, as an allegory of the mechanism itself. This results in an epistemic insight: Humans themselves, when performing a mathematical calculation or carrying out logical reasoning with pencil on paper, are already in a momentary machine state,[14] and therefore temporarily become an algorithmic 'it' rather than a gendered 'he' or 'she'. In the computational space that unfolds between the human user and the AI-generated results, multiple software and hardware layers of artificial neural nets intervene, turning the apparent dialogue with the machine into a more intransitive dia(techno)logue.

But there are ways to make this process transparent again. Once the 'Play sketch' button is activated, the Artificial Neuron Trainer inputs in a real-time editor allow almost behaviouristically (as stimulus-response formation) the reading back of any change in the output as a function of alterations in the source code that is revealed in another window on the same browser screen.[15] Understanding occurs here both from the human and the machine perspective.[16] A different kind of second-order transitivity occurs in *thinkering* with contemporary AI tools. Hands-on programmed examples invite a conscious re-enacting of what occurs behind the curtains of machine-learning applications like Keras that have been 'designed for human beings, not machines',[17] and other open-source AI program libraries like PyTorch.[18] Their mechanism invites for real-time epistemic reflection. Still, even the live coding aesthetics for AI applications, watching changing results as immediate functions of parameter changes in the source code, rather dissimulates the precise internal media processes that unfold in between. The concrete computing operations, in terms of machine language and hardware, still remain opaque and therefore allow for transitive communication only indirectly. The techno-mathematical mechanism of machine learning can no longer be sufficiently analysed by reducing it inductively down to its single elements such as artificial neurons. It deals with statistical probabilities and approximations instead. The behaviour of algorithms in combination with big data cannot be finitely predicted (in accordance with Kurt Gödel's theorem of mathematical computation). In reverse, it is not possible to sufficiently deduce from AI-generated output the chain of transformations of input that have been occurring within a

14. Turing, 'On Computable Numbers', p. 231.
15. In the experimental online packet accompanying the print version of Noack and Sanner, *Künstliche Intelligenz verstehen*, corresponding with chapter 12, 'Neuronale Netze III: Fehler zurückverfolgen mit dem Neuronentrainer'.
16. https://editor.p5js.org/MaschinenNah/sketches/pPH-1PFOT
17. *Keras*, https://keras.io [accessed 27 August 2024].
18. Noack and Sanner, *Künstliche Intelligenz verstehen*, p. 26.

trained artificial neural net, as in generative adversarial networks. Such an aim requires hypothetical or abductive reasoning instead, which 'merely suggests that something may be'.[19]

Stochastic processuality no longer entails a direct object; its understanding therefore becomes intransitive in the grammatical sense. Experimenting with diagrammatic models shifts critical *analysis* from the immediacy of hardware to algebraic formulas. 'Loosely coupled' artificial neural nets (in the *medium* sense of Fritz Heider[20]) act *synthetically*. In techno-philosophical terms they undo the hardware vs software dialectics. They remind, as a kind of technological mirror, that even the human mind has never been an abstract representation *of*, but has always been entangled *with*, and enacted *by*, the physical world—be it biochemical, electronic or otherwise. Quantum physics knows the collapse of the object–observer distance from the technical setting of a measuring apparatus in relation to its nanophysical object. From the point of view of artificial intelligence, there has never been an *ontological* dichotomy between spirit and matter, software and hardware—even if these differences are still *operationally* maintained. The categorial gap between a thought model and its mate*realization* as technical device has already been transformed by the Turing machine, with its medium message being a mechanization of mathematics rather than a simply logified hardware. Essentially, the technical thing hereby becomes identical with techno-logical thought itself. But the technical modelling of biochemical neurons via interconnected binary-switching elements into an artificial neural network (starting in early AI with electronic tubes and finally resulting in integrated circuits) still requires a clarification concerning the electronic and the organic embodiment of thought: 'We are not comparing the structure of neurons with the structure of transistors; we are comparing the organization of symbol-processing procedures in an algorithm with symbol-processing procedures of the mind-brain.'[21]

Pioneer of computational neurophysiology Walter Rosenblith questioned the elementarization, reductionism and archaization in artificial intelligence already in times of still-tangible electronic units of operation like the trigger relay: 'We no longer hold the earlier widespread belief that the so-called all-or-none-law from nerve impulses makes it legitimate to think of relays as adequate models for neurons. [...] In a system whose numerous elements interact so strongly with each other, the functioning of the system is not necessarily best understood by proceeding on a neuron-by-neuron basis *as if*

19. Peirce, 'Lectures on Programatism', p. 171.
20. As defined in Heider, 'Thing and Medium', pp. 1–35.
21. Colby, *Artificial Paranoia*, p. 25.

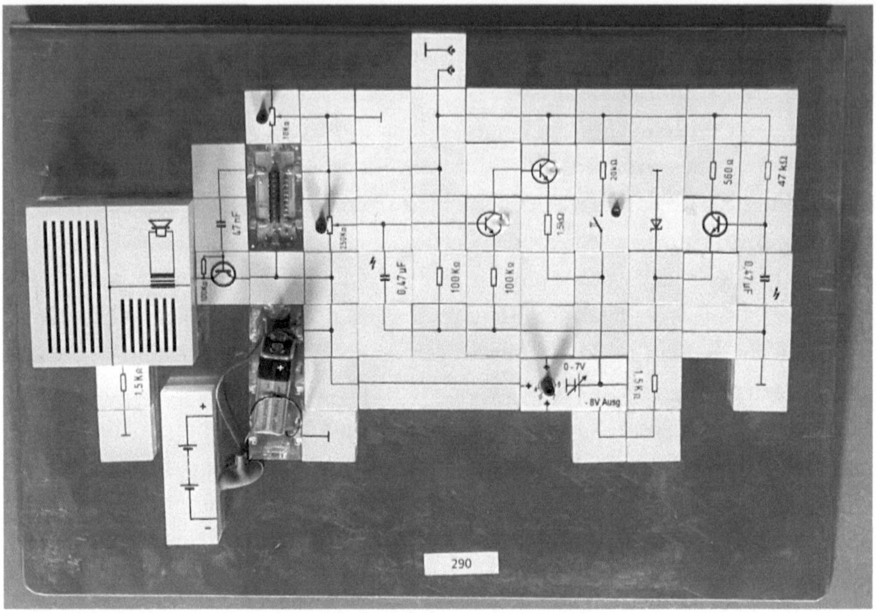

Figure 4.1: Artificial neuron model from the Lectron 'Neurophysiology Experiments' construction kit.

each had an independent personality'.[22] And in times of high-frequency parallel computing, it would take more than a human lifetime for the realization of such a stepwise analysis. Any understanding of emerging patterns would be literally dissipated over time. Does that mean that the categorical claim for a 'transitive' philosophy of complex non-linear and stochastic pattern generation is inappropriate as such, just as from the final state of a chess game the exact genealogy of its previous single actions cannot logically be traced back?

Contemporary convolutional neural nets are not simply processing data symbolically, like (rather language- than machine-oriented) software in the familiar way that remained, at least conceptually, largely intransitive towards its physical implementation. Much more subtly than binary 'information' that is somewhat violently imposed upon brute matter, CNNs rather subsymbolically operate by signal transduction from within techno-mathematically (non-linear algebra) organized matter itself. With 'deep' artificial neural nets and their (albeit still numerically) 'weighted' neural action potentials the change of the massively data-trained output is still a function, but not

22. Walter A. Rosenblith, 'On Cybernetics and the Human Brain', *The American Scholar* (Spring 1966), p. 247, as quoted in: Dreyfus, *What Computers Still Can't Do*, p. 162 (emphasis added).

completely proportional to the change of the input. Media transitivity has thereby become self-acting while at the same time challenging direct human media insight that still aims at 'entering the instrument' with techno-philosophical intention.[23] Thus, a challenge for operative media-artefactual 'labs' like the Media Archaeological Fundus arises: Does the media-epistemic momentum of AI escape its haptic experience, in favour of a mathematization that can neither be directly cognized nor be displayed materially for transitive inspection any longer? A more 'radical' (square root) archaeology of computational media knowledge is therefore required. The core of contemporary AI is still hardware and software, but its media-epistemic computing power derives from its shift towards conceptual diagrammatics and algebraic modelling which intervenes (in a more abstract 'medium' sense) against transitive human cognition that tries to analytically reflect the techno-logical generation of emerging 'intelligent' effects. AI unfolds in an in/transitive double-bind. In a physical sense, of course, the requirements of massive data training still relate machine learning transitively to 'power' but in a less discursive, rather energetic and material sense of electricity, temperature cooling and microchip capacity. This becomes infrastructurally concrete in so-called server farms.

Is the human observer thereby finally excluded from direct insight into the technological events of AI and condemned to stay at media-theoretical distance, in favour of an inner-technical media reasoning by artificial neural nets themselves? There is indeed an 'algorithmic reason as a non-linear elaboration of continuous infinities and transformation of its discrete parts'.[24] According to Turing, the time-discrete machine, at every given moment, is 'directly aware' (if not conscious) of its configurations. There is cognition in the system.[25] *Self*-consciousness though, as a condition for the epistemological reflection of technical processes in real-time, according to Gotthard Günther's 'transcendental cybernetics' (1963), requires a metalanguage that imposes intransitivity again. Contrary to the cybernetic system approach, it is the material gap between electronic and organic instantiation and embodiment of signal processing that still sets limits to AI's becoming epistemically self-conscious in a cognitively trained neural net called the human brain. A transitive media philosophy of AI remains an option only as a hybrid solution. But first of all this requires the human to admit the artificiality—that is, the technicity of *both* embodied and electronic reasoning—in a dynamic update of the very term technology.

23. Luciana Parisi in Panayotov, '"To Engineer the Time by Other Means"'.
24. Parisi, 'Automated Architecture', p. 413.
25. Turing, 'On Computable Numbers', p. 231.

In 'deep' machine learning, classical hermeneutics itself (the art of decoding the several levels of meaning in textual expressions) has become operational, with its multiple layers of artificial neuron configurations that are transforming input signals into outputs. There is an implicit techno-logic of stochastic probabilities and emerging patterns from noise that is not simply burnt into the microchips of the graphics processing unit in its logical combinatorial circuits. Complex neural nets, in their inner states, are rather time-dependent as a function of their short- or long-term big data 'memories'. Transparency does not derive from a direct correlation of output to input here, but becomes a function of non-linear signal processing. AI's inherent knowledge therefore demands to be epistemically revealed as media intelligence *while* it techno-logically occurs in the media act.

Experiencing Artificial Imaging Intelligence Transitively

Technological transparity is no longer simply resulting from technical literacy in computation, as it used to in classic computing culture. The critical task of 'getting inside the algorithms themselves' becomes more complex once human-scale reasoning is confronted with algorithms that are altered as they 'learn' on training data.[26] One and the same human textual prompt input can lead to different outputs in AI tools such as the open-source text-to-image generator Stable Diffusion and its extension to VideoFusion. Changing a starting value (seed) for a pseudo-random generator results in non-linear transformations that occur within the artificial neural net. Its inner reasoning invites an 'applied speculative realism' in the effort of a signal-based enquiry into the inner-technical media theatre of AI-generated aesthetics.[27] For a media-reflexively controlled experimentation (rather than simply phenomenological trial and error) not only the source code but the concrete operations in the 'hidden layers' of artificial neural nets are expected to become transparent.[28] Such a practice consciously turns classical technical *tinkering* into machine-conscious *thinkering*. A transitive media AIsthetics does not get lost in the vast ocean of the techno-mathematical ensemble. The media-philosophical 't(h)inkerer' does not remain a distanced observer or descriptor of the technical AI object but necessarily becomes cognitively entangled with the artificial neural mechanism itself. The human does not only get analytically involved in the engineering circuitry as agency of positive or negative signal

26. Burrell, 'How the Machine "Thinks"', p. 2.
27. Bogost, *Alien Phenomenology*, p. 29.
28. Rombach et al., 'High-Resolution Image Synthesis with Latent Diffusion Models'.

feedback but, in return, brings the machine 'into the loop' of their techno-philosophical insights.

While big data computing as machine learning may generate visual shapes from noisy signals in the diffusion model, it still requires human recognition to reveal its latent 'log.icon' behind the visual human–machine interface.[29] In search of an epistemic surplus, the focus on decision-critical details allows for reducing the computational complexity of AI image generation into precise, almost transitive explainability: at least by exemplary sampling of scenarios that metonymically stand for the overall event. A verbal, algebraic and diagrammatic case analysis of AI image processing may come close to the technical event but is still reductive: It refers to the symbolic level of the machine, not to its actual implementation. The notion of visual 'abstraction' demands to be redefined here, since it is no indexical image that is mathematized; instead an already digitized, numerical (still physically electric current) value is transformed into another mathematical configuration. The very notion of representation is questioned by the concrete technological operation and is revealed as an anthropocentric metaphor of brain action.

The processual idiosyncrasies and techno-logics of AI image generation cannot directly be deduced from the final results on a computer monitor. The precise identification of its moments is therefore the task of *explainable* AI tools such as the CLIP system, which classifies text–image pairings in a data set. Via ambiguous prompts, polysemies can lead to significant image differences. But it is exactly from such 'mis'-interpretations on the machine side that a true *media understanding* results. Once such 'errors' are epistemically granted the status of a techno-logical articulation, they become concrete instantiations of 'epistemic porosity', where technical knowledge and philosophical oversight 'must somehow coexist and interpenetrate'.[30] The conjunction of text and image, within an artificial neural net, can symbolically be expressed by a matrix. But what looks like a familiar mathematical tool can be more fundamentally interpreted as a deconstruction of the cultural text/image-difference.[31] Even more challenging is the radical questioning of iconology at all, when the visual order arises from aniconic noise in the re-emergence of pixel 'images' from latent space. Mimicking statistical physics as known from thermodynamics, such a computational process implements equally distributed disorder into the virtual—that is, techno-mathematically

29. A media-archaeologically stimulating neographism by Striano, 'Log.icona', pp. 77–95.
30. In a slightly modified sense of Berry, 'Porosity and Computation'.
31. For the pre-technical insistence on the medium specificity of text vs image in aesthetics, see Lessing, *Laocoon*.

calculating—space of AI. Initially an input image is gradually transformed into Gaussian noise by means of Markov chains.[32] All of a sudden, a cultural artefact like the 'image' becomes redefined in terms of physical entropy, as has already been recognized in the heroic epoch of cybernetic information aesthetics.[33] The more detailed its techno-mathematical description becomes, the more acute may be its hidden epistemic relevance. But it takes the media-philosophically attentive mind to co-articulate this technically hidden knowledge and to reveal its inherent epistemic beauty.

The mismatch between 'deep' machine learning (such as Q-Learning, a form of reinforcing qualities in artificial neural nets) and human cognition still results from a relative lack: the long-time experience of an environmental world that can only be approximated by big data training models. But such a machine deficiency can be media-archaeologically considered as a strength, since it allows for techno-logical autopoiesis. When in similarity-based image retrieval the topic of the human hand is, next to more plausible results, suddenly identified with the image of a pig, the apparent 'error' may turn out, in a more benevolent techno-epistemic cognition, to be an indication of the alternative logic of machine vision: it parses images into essential features like distinct limbs that in their numerical weight result in such surreal configurations.[34] In human vs machine-learning games, advanced subsymbolic artificial intelligence has no *a priori* knowledge about the rules or the games but induces them from the successive states of the gaming procedures. Media-epistemologically, the issue of a *tec-knowledge without understanding* arises, as has been already discussed in the debate about 'strong' vs 'weak' symbolical intelligence.[35]

32. See Sohl-Dickstein et al., 'Deep Unsupervised Learning Using Nonequilibrium Thermodynamics'. The author derived much of this information and references from Raphael Johannes Tostlebe's practice-based and prize-winning master thesis *Die Dramaturgie KI-generierter Videos: Zwischen Anthropozentrismus und Technológos* (Department of Media Science, Humboldt University of Berlin, April 2024).

33. As became prominent with Moles, *Théorie de l'information et perception esthétique*.

34. On more idiosyncrasies in automated image retrieval, see Ernst, Heidenreich and Holl (eds), *Suchbilder*. See also Ernst, 'There Is No "Error" in Techno-Logics', pp. 79–94.

35. See especially Searle's 'Chinese Room' thought experiment: Searle, 'Minds, Brains, and Programs', pp. 417–57.

Epilogue

Media-Active Philosophy

Techno-logically (in-)formed matter, as physical embodiments of engineering and mathematical knowledge, is intended to act rational. Still, such techno-logical media artefacts are not simply retroactive to, or mirroring, those human programming designs that they materialize. Their algorithmic unpredictabilities, technological idiosyncrasies and physical frictions articulate an implicit surplus tec-knowledge of their own. Such a fact may appear familiar from asymmetrical, critical or even non-linear human–nature interaction as well, as is currently experienced in the environmental climate crisis. But in contrast to the co-called Anthropocene as the current geological era where cultural intervention has become critically invasive to the Earth's energy balance, the Technologo(s)cene, in spite of all its physical detriments, remains an enaction of a kind of mind. In times of a not just human (media) culture, software-driven machines have begun to develop a 'love for *lógos*' themselves.[1] Technical mechanisms, as material objectifications of reasoning, have so far been dependent on human reflection to be explicitly expressed as epistemic tools. But from the moment when Boolean false/true logic was identified as isomorphic with networks of electromagnetic relays, philosophy *of* technology has been extended towards a post-human media-active philosophy *as* technology. Such a media thinking in the medio-passive voice bridges the formerly epistemic dichotomy and transcends the gap that has separated humans and machines so far, granting transitivity to each other's insight both ways. Tightly coupled to the medium, the transitive philosopher becomes medium itself. It is up to the reader (human or non-human) to speculate whether this thesis has been generated and formulated from a customized AI chatbot that has been trained by the big data archive of media-epistemic texts (including the present treatise). Such an intelligence, once it has been 'deeply' integrated in the loops of circuits and their manifold layers (electronic or otherwise), may result in an inner-technical transitive media philosophy that is becoming self-conscious *of* and *as* the machine itself.

1. A strong argument in Hiller's PhD thesis, *Maschinenphilologie*.

Such are the kind of thoughts that transitive media philosophy is obsessed with. Its 'transitivity' claims more than just a retroactive nostalgia for tangible tec-knowledge in times of increasingly black-boxed technologies. Its methodic specificity rather arises from the claim for an instant epistemic awareness of technical practices in their operative presence, be they physically material or techno-mathematically abstracted. As has been repeatedly insisted in the preceding arguments, the philosophical moment(um) of transitive media insight, complementary to a more distanced media theory and its textuality, takes place when a technical device is in its actual signal-processing state. This requires a direct cognitive coupling of epistemological reason (a philosophically attuned and technical knowledge-replete mind) to the diagrammatic reasoning of the mechanism itself to immediately experience the occasional media event as invitation for epistemic insight. The condition for such a techno-philosophical alertness is that the respective media operations are perceived as coextensive to the thoughts they induce in real-time presence. A most transitive media-philosophical experience, therefore—and very ironically—is the abrupt finitude of this very argument through the switching off of its computational word processing.

Bibliography

Note: Years marked by asterisks in brackets [*] indicate the original publication; "GO" and "FO" indicate the original year of publication in German or French for works that have been translated into English

Abraham, Ralph and Yoshisuke Ueda, *The Chaos Avant-Garde: Memories of the Early Days of Chaos Theory* (Singapore: World Scientific, 2000). https://doi.org/10.1142/9789812386472

Agamben, Giorgio, 'A Philosophy of Contact', transl. by Richard Braude, *Ill Will* (entry 11 January 2021). https://illwill.com/philosophy-of-contact

Angerer, Marie-Luise, *Nonconscious: On the Affective Synching of Mind and Machine* (Lüneburg: Meson Press, 2022)

Angerer, Marie-Luise, Michaela Ott and Bernd Bösel (eds), *Timing of Affect: Epistemologies, Aesthetics, Poltics* (Zurich and Berlin: diaphanes, 2014)

Bachelard, Gaston, *Études* (Paris: Librairie philosophique J. Vrin, 1970)

Barad, Karen, *Meeting the Universe Halfway: Quantum Physics and the Entanglement of Matter and Meaning* (Durham and London: Duke University Press, 2007). https://doi.org/10.2307/j.ctv12101zq

Barthes, Roland, 'Leaving the Movie Theater', in idem, *The Rustle of Language*, transl. by Richard Howard (New York: Hill and Wang, 1986), pp. 345–49

Barthes, Roland, 'Myth Today', in idem, *Mythologies* [FO 1957], selected and transl. by Annette Lavers (New York: Noonday Press, 1991), pp. 107–64

Barthes, Roland, 'To Write: An Intransitive Verb?' [FO 1966], in idem, *The Structuralist Controversy: The Language of Criticism and the Sciences of Man*, ed. by Richard Macksey and Eugenio Donato (Baltimore: Johns Hopkins University Press, 2007), pp. 134–45

Bense, Max, 'The Projects of Generative Aesthetics', in Jasia Reichardt (ed.), *Cybernetics: Art and Ideas* (London: Studio Vista, 1971), pp. 57–60

Bergson, Henri, *Creative Evolution* [1911] (Mineola, NY: Dover, 1998)

Berry, David M., 'Tracing "Toxicity" Through Code: Towards a Method of Explainability and Interpretability in Software', *Digital Humanities Quarterly* (*DHQ*), 17, no. 2 (2023). https://orcid.org/0000-0002-7737-5586

Berry, David M., 'Porosity and Computation', *stunlaw* (December 2024). http://stunlaw.blogspot.com/2024/12/porosity-and-computation.html [accessed 14 December 2024]

Bjørn, Kim, *Patch & Tweak with Moog* (Denmark: Bjooks, 2nd edn 2021)

Bogost, Ian, *Alien Phenomenology, Or: What It's Like to Be a Thing* (Minneapolis and London: University of Minnesota Press, 2012). https://doi.org/10.5749/minnesota/9780816678976.001.0001

Bolter, J. David, *Turing's Man: Western Culture in the Computer Age* (Chapel Hill: University of North Carolina Press, 1984)

Borbach, Christoph, *Delay: Mediengeschichten der Verzögerung, 1850–1950* (Bielefeld: transcript, 2024). https://doi.org/10.1515/9783839452349

Boros, Lili, 'Visual and Textual Narratives in the Photofilms of Gusztáv Hámos and Katja Pratschke', in Hámos, Gusztáv and Katja Pratschke, *In Limbo* (Berlin: Revolver Publishing, 2024), pp. 221–25

Brecht, Bertolt, 'Der Rundfunk als Kommunikationsapparat: Rede über die Funktion des Rundfunks', *Blätter des Hessischen Landestheaters Darmstadt*, 16 (July 1932)

Bredekamp, Horst, *Image Acts: A Systematic Approach to Visual Agency* (Berlin: De Gruyter, 2018). https://doi.org/10.1515/9783110548570

Bryant, Levi R., *Onto-Cartography: An Ontology of Machines and Media* (Edinburgh: Edinburgh University Press, 2014). https://doi.org/10.1515/9780748679980

Burrell, Jenna, 'How the Machine "Thinks": Understanding Opacity in Machine Learning Algorithms', *Big Data & Society*, 3, no. 1 (2016), pp. 1–12. https://doi.org/10.1177/2053951715622512

Canales, Jimena, *A Tenth of a Second: A History* (Chicago and London: University of Chicago Press, 2009)

Carpenter, Edmund and Marshall McLuhan, 'Acoustic Space', in idem (eds), *Explorations in Communication: An Anthology* (Boston: Beacon Press, 1960), pp. 67–69

Carter, B., 'Large Number Coincidences and the Anthropic Principle in Cosmology', in M.S. Longair (ed.), *Confrontation of Cosmological Theories with Observational Data* (Dordrecht and Boston: Reidel, 1974), pp. 291–98. https://doi.org/10.1007/978–94-010-2220-0_25

Ceruzzi, Paul E., 'The Early Computers of Konrad Zuse, 1935–1945', *Annals of the History of Computing*, 3, no. 3 (July 1981), pp. 241–62. https://doi.org/10.1109/MAHC.1981.10034

Clark, Andy and David Chalmers, 'The Extended Mind', *Analysis*, 58, no. 1 (1998), pp. 7–19. https://doi.org/10.1093/analys/58.1.7

Colby, Kenneth M., *Artificial Paranoia: A Computer Simulation of Paranoid Processes* (Elmsford, NY: Pergamon Press, 1975)

Conrad, Michael, 'The Prize of Programmability', in Rolf Herken (ed.), *The Universal Turing Machine: A Half-Century Survey* (Vienna and New York: Springer, 2nd edn 1994), pp. 261–81. https://doi.org/10.1007/978-3-7091-6597-3_11

Craven, T.L., 'An Engineering Application of Boolean Algebra', *Proceedings of the Royal Irish Academy*, 57, no. 6 (1955), pp. 121–30

Debray, Regis, *Media Manifestos: On the Technological Transmission of Cultural Forms* (London and New York: Verso Books, 1996)

DeLanda, Manuel, *Intensive Science and Virtual Philosophy* (London: Continuum, 2002)

Deleuze, Gilles and Michel Foucault, 'Intellectuals and Power', in Michel Foucault, *Language, Counter-Memory, Practice: Selected Essays and Interviews*, ed. by D.F. Bouchard (Ithaca, NY: Cornell University Press, 1977), pp. 205–17

Dennhardt, Robert, *Die Flipflop-Legende und das Digitale: Eine Vorgeschichte des Digitalcomputers vom Unterbrecherkontakt zur Röhrenelektronik 1837–1945* (Berlin: Kulturverlag Kadmos, 2009)

Dolar, Mladen, *A Voice and Nothing More* (Cambridge, MA: MIT Press, 2006). https://doi.org/10.7551/mitpress/7137.001.0001

Dörfling, Christina, *Der Schwingkreis: Schaltungsgeschichten an den Rändern von Musik und Medien* (Munich: Fink, 2022). https://doi.org/10.30965/9783846766859

Dreyfus, Hubert L., *What Computers Still Can't Do: A Critique of Artificial Reason* (Cambridge, MA and London: MIT Press, 1994)

Eccles, W.H. and F.W. Jordan, 'A Trigger Relay Utilising Three-Electrode Thermionic Vacuum Tubes', *Radio Review*, 1 (December 1919), pp. 143–46

Enders, Bernd, 'Mathematik ist Musik für den Verstand, Musik ist Mathematik für die Seele. Zum Verhältnis von Ton und Zahl, Klang und Gefühl, Musik und Technik, Sound und Computer', in idem (ed.), *Mathematische Musik—musikalische Mathematik* (Saarbrücken: Pfau, 2005), pp. 7–37

Ernst, Wolfgang, *Sonic Time Machines: Explicit Sound, Sirenic Voices and Implicit Sonicity in Terms of Media Knowledge* (Amsterdam: Amsterdam University Press, 2016). https://doi.org/10.1017/9789048528479

Ernst, Wolfgang, 'There Is No "Error" in Techno-Logics: A Radically Media-Archaeological Approach', in Maria Korolkova and Timothy Barker (eds), *Miscommunications: Errors, Mistakes, Media* (London: Bloomsbury Academic, 2021), pp. 79–94. https://doi.org/10.5040/9781501363825.0011

Ernst, Wolfgang, *Technológos in Being: Radical Media Archaeology & the Computational Machine* (New York: Bloomsbury Academic, 2021). https://doi.org/10.5040/9781501362262

Ernst, Wolfgang, 'On the Notion of Cybernetic Frictions and Its Role in the Deployment of a Radical Media Archaeology', in Diego Gómez-Venegas (ed.), *Frictions: Between Cybernetic Thinking and Its Attempts Towards Mate[real]ization* (Lüneburg: Meson Press, 2023), pp. 119–30

Ernst, Wolfgang, Stefan Heidenreich and Ute Holl (eds), *Suchbilder: Visuelle Kultur zwischen Algorithmen und Archiven* (Berlin: Kulturverlag Kadmos, 2003)

Evens, Aden, 'Object-Oriented Ontology, or: Programming's Creative Fold', *Angelaki: Journal of the Theoretical Humanities*, 2, no. 1 (April 2006), pp. 89–97. https://doi.org/10.1080/09697250600797930

Fazi, M. Beatrice, *Contingent Computation: Abstraction, Experience, and Indeterminacy in Computational Aesthetics* (London: Rowman & Littlefield, 2018). https://doi.org/10.5040/9798881812027

Feld, Steven, 'Acoustemology', in David Novak and Matt Sakakeeny (eds), *Keywords in Sound* (Durham, NC: Duke University Press, 2015), pp. 12–21. https://doi.org/10.2307/j.ctv11sn6t9.4

Fischer, Thomas and Andrei Cretu, 'The Ashby Box: A Contextualization and Speculative Remake', in Diego Gómez-Venegas (ed.), *Frictions: Between Cybernetic Thinking and Its Attempts Towards Mate[real]ization* (Lüneburg: Meson Press, 2023), pp. 163–93

Foerster, Heinz von, *Understanding Understanding: Essays on Cybernetics and Cognition* (New York: Springer, 2003)

Foucault, Michel, 'The Statement and the Archive', in idem, *The Archaeology of Knowledge and the Discourse on Language* [FO 1969], transl. by A.M. Sheridan Smith (London and New York: Pantheon Books, 1972), pp. 77–131

Franke, Herbert W., *Computergraphik—Computerkunst* [*1971] (Heidelberg: Springer, 2nd ext. edn 1985). https://doi.org/10.1007/978-3-662-06388-0

Franke, Herbert W. and Gottfried Jäger, *Apparative Kunst* (Cologne: DuMont, 1973)

Frommolt, Karl-Heinz and Martin Carlé, 'The Song of the Sirens', *The Nordic Journal of Aesthetics*, 48 (2014), pp. 18–33

Gandorfer, Daniela and Zulaikha Ayub, 'Introduction: Matterphorical', *Theory & Event Journal* special issue, 24, no. 1 (2021). https://doi.org/10.1353/tae.2021.0001

Gardner, Martin, *Logic Machines and Diagrams* (New York: McGraw-Hill, 1958)

Gibson, James J., 'The Theory of Affordances', in Robert Shaw and John Bransford (eds), *Perceiving, Acting and Knowing* (New York: John Wiley, 1977), pp. 67–82

Giessmann, Sebastian, *The Connectivity of Things: Network Cultures since 1832* (Cambridge, MA and London: MIT Press, 2024). https://doi.org/10.7551/mitpress/12547.001.0001

Günther, Gotthard, 'Die "zweite" Maschine' ['The "Second" Machine'], in idem, *Das Bewußtsein der Maschinen. Eine Metaphysik der Kybernetik* (Krefeld and Baden-Baden: Agis, 2nd edn 1963), Supplement IV, pp. 179–203

Haberer, Maximilian, 'Phonographendämmerung? Oder: Wie man mit dem Tonband philosophiert', in idem, *Tape Matters: Ästhetik, Materialität und Klangkonzepten des Tonbandes* (Düsseldorf: Düsseldorf University Press, 2023), pp. 265–77. https://doi.org/10.1515/9783111453064-005

Harman, Graham, *Tool-Being: Heidegger and the Metaphysics of Objects* (Peru, IL: Open Court, 2002)

Hayles, N. Katherine, *How We Became Posthuman: Virtual Bodies in Cybernetics, Literature, And Informatics* (Chicago and London: University of Chicago Press, 1999). https://doi.org/10.7208/chicago/9780226321394.001.0001

Hayles, N. Katherine, 'Print Is Flat, Code Is Deep: The Importance of Media-Specific Analysis', *Poetics Today*, 25, no. 1 (2004), pp. 67–90. https://doi.org/10.1215/03335372-25-1-67

Hayles, N. Katherine, *How We Think: Digital Media and Contemporary Technogenesis* (Chicago and London: University of Chicago Press, 2012). https://doi.org/10.7208/chicago/9780226321370.001.0001

Hayles, N. Katherine, *Unthought: The Power of the Cognitive Nonconscious* (Chicago and London: University of Chicago Press, 2017). https://doi.org/10.7208/chicago/9780226447919.001.0001

Hegel, Georg Wilhelm Friedrich, *Phenomenology of Spirit* [GO 1807] (Oxford: Oxford University Press, 1977)

Heidegger, Martin, *Sein und Zeit* [GO 1927] (Tübingen: Max Niemeyer, 1993)

Heidegger, Martin, 'The Question Concerning Technology' [GO 1954], in idem, *Basic Writings*, ed. by David Farrell Krell (New York: Harper & Row, 1977), pp. 287–317

Heider, Fritz, 'Thing and Medium' [GO 1927], in idem, *On Perception, Event Structure, and Psychological Environment: Selected Papers* (New York: International Universities Press, 1959], pp. 1–35

Helmholtz, Hermann von, *On the Sensations of Tone as a Physiological Basis for the Theory of Music* [GO 1863], transl. by Alexander J. Ellis (London: Longmans Green, 1895)

Hertz, Garnet and Jussi Parikka, 'Zombie Media: Circuit Bending Media Archaeology into an Art Method', *Leonardo*, 45, no. 5 (2012), pp. 424–30. https://doi.org/10.1162/LEON_a_00438

Herzogenrath, Bernd, '"Wait a Minute": Reflections on [the] Pause', *MLN* (*Modern Language Notes*), 137 (2022), pp. 1031–41. https://doi.org/10.1353/mln.2022.0074

Higinbotham, W.A., 'The Model 200 Pulse Counter', *Review of Scientific Instruments*, 18 (October 1947). https://doi.org/10.1063/1.1740832

Hill, Kaycee, 'Creating a Website with Websim AI: Here's How it Works', *tom's guide* [published 9 July 2024]. https://www.tomsguide.com/how-to-use-websim [accessed 12 July 2024]

Hiller, Moritz, *Maschinenphilologie* (Berlin: Kulturverlag Kadmos, 2023)

Hiller, Moritz and Stefan Höltgen (eds), *Archäographien: Aspekte einer Radikalen Medienarchäologie* (Berlin: Schwabe, 2019). https://doi.org/10.31267/978-3-7574-0029-3

Horkheimer, Max, 'Traditional and Critical Theory' [GO 1937], in idem, *Critical Theory: Selected Essays* (New York: Continuum, 1982), pp. 188–243

Huhtamo, Erkki, 'Thinkering with Media: On the Art of Paul DeMarinis', in Ingrid Beirer, Sabine Himmelsbach and Carsten Seiffarth (eds), *Paul DeMarinis: Buried in Noise* (Heidelberg and Berlin: Kehrer, 2010), pp. 33–46

Hui, Yuk, 'Deduktion, Induktion und Transduktion: Über Medienästhetik und digitale Objekte', *Zeitschrift für Medienwissenschaft*, 8, no. 1 (2013), pp. 101–15

Hui, Yuk, *The Question Concerning Technology in China: An Essay in Cosmotechnics* (Falmouth: Urbanomic, 2016)

Husserl, Edmund, *On the Phenomenology of the Consciousness of Internal Time* [*1928], transl. by J.B. Brough (Dordrecht: Kluwer Academic Publishers, 1991)

Husserl, Edmund, *The Crisis of the European Sciences and Transcendental Phenomenologie* (Evanston, IL: Northwestern University Press, 1970)

Jakubaschk, Hagen and Ludwig Scholz, *Fernsehempfänger selbstgebaut* (Neuenhagen: Sport und Technik, 1960)

Johnson, Danielle 'Danz', 'Alive and Analog' (interview), in Kim Bjørn, *Patch & Tweak with Moog* (Denmark: Bjooks, 2nd edn 2021)

Kahn, Douglas, *Earth Sound Earth Signal: Energies and Earth Magnitude in the Arts* (Berkeley, CA: University of California Press, 2013). https://doi.org/10.1525/9780520956834

Kaiser, Mario, 'Do You Speak Technology? On Grammars of Technology', in Gerhard Gamm et al. (eds), *Jahrbuch Technikphilosophie 2015* (Berlin: diaphanes, 2015), pp. 117–36

Kittler, Friedrich A., 'There Is No Software', in idem, *Literature, Media, Information Systems: Essays* (London: Routledge, 1997), pp. 147–55

Kittler, Friedrich A., *Gramophone, Film, Typewriter* (Stanford: Stanford University Press, 1999)

Kittler, Friedrich A., 'Protected Mode', in idem, *The Truth of the Technological World* (Redwood City: Stanford University Press, 2020), pp. 209–18. https://doi.org/10.1515/9780804792622-015

Kolb, Richard, *Das Horoskop des Hörspiels* (Berlin: Max Hesse, 1932)

Koopman, Phil, 'Microcodes versus Hard-Wired Control: A Comparison of Two Methods for Implementing the Control Logic for a Simple CPU', *BYTE* (January 1987), pp. 235–42

Kracauer, Siegfried, *Theory of Film: The Redemption of Physical Reality* [*1960] (Princeton, NJ: Princeton University Press, 1997)

Krämer, Sybille, 'Geist ohne Bewußtsein? Über einen Wandel in den Theorien vom Geist', in idem (ed.), *Geist, Gehirn, Künstliche Intelligenz: Zeitgenössische Modelle des Denken* (Berlin and New York: De Gruyter, 1994), pp. 88–110. https://doi.org/10.1515/9783110850505

Landauer, Rolf , 'Irreversibility and Heat Generation in the Computing Process', *IBM Journal of Research and Development*, 5, no. 3 (1961), pp. 183–91. https://doi.org/10.1147/rd.53.0183

Landauer, Rolf, 'Information is Physical', *Physics Today*, 44, no. 5 (1991), pp. 23–29. https://doi.org/10.1063/1.881299

Laruelle, François, 'L'ordinateur transcendantale: une utopie non-philosophique', in idem (ed.), *Homo ex machina* (Paris: L'Harmattan, 2005), pp. 89–150

Lessing, Gotthold Ephraim, *Laocoon: An Essay upon the Limits of Painting and Poetry* [GO 1766] (Cambridge: Cambridge University Press, 2012)

Libet, Benjamin, 'Unconscious Cerebral Initiative and the Role of Conscious Will in Voluntary Action', *Behavioral and Brain Sciences*, 8, no. 4 (December 1985), pp. 529–39. https://doi.org/10.1017/S0140525X00044903

Licklider, Joseph Carl Robnett, 'Man–Computer Symbiosis', in *IRE Transactions on Human Factors in Electronics*, 1, no. 1 (March 1960), pp. 4–10. https://doi.org/10.1109/THFE2.1960.4503259

Lohmann, Johannes, *Musiké und Logos* (Stuttgart: Musikwissenschaftliche Verlags-Gesellschaft, 1970)

Longo, Giuseppe and Jean Lassègue, 'Actualité de Turing: entre captation d'héritage et ressource pour l'avenir', *Intellectica*, 72 (2020), pp. 215–36. https://doi.org/10.3406/intel.2020.1953

Lord, Albert B., *The Singer of Tales* (Cambridge, MA: Harvard University Press, 1960)

Lorenz, Edward, 'The Butterfly Effect', in Ralph Abraham and and Yoshisuke Ueda (eds), *The Chaos Avant-Garde: Memories of the Early Days of Chaos Theory* (Singapore: World Scientific, 2000), pp. 91–94. https://doi.org/10.1142/9789812386472_0007

Lucier, Alvin, 'Origins of a Form: Acoustical Exploration, Science and Incessancy', *Leonardo Music Journal*, 8 (1998), pp. 5–11. https://doi.org/10.2307/1513391

Maeda, Yoshinobu, and Hideo Makino, 'A Pulse-Type Hardware Neuron Model with Beating, Bursting Excitation and Plateau Potential', *BioSystems*, 58 (2000), pp. 93–100. https://doi.org/10.1016/S0303-2647(00)00111-8

Manning, Erin and Brian Massumi, *Thought in the Act: Passages in the Ecology of Experience* (Minneapolis, MN and London: University of Minnesota Press, 2014). https://doi.org/10.5749/minnesota/9780816679669.001.0001

Marks, Laura U., 'How Electrons Remember', *Millennium Film Journal*, 34 (Fall 1999), thematic issue 'The Digital'. http://www.mfj-online.org/journalPages/MFJ34/LMarks.html [accessed 11 January 2025]

Marquand, Allan, 'A New Logical Machine', *Proceedings of the American Academy of Arts and Sciences*, 21 (1885/6), pp. 303–07. https://doi.org/10.2307/25129824

Massumi, Brian, 'The Autonomy of Affect', *Cultural Critique*, 31 (September 1995), pp. 83–110. https://doi.org/10.2307/1354446

Massumi, Brian, 'The Crannies of the Present', in Stuart Grant, Jodie McNeilly, and Maeva Veerapen (eds), *Performance and Temporalisation: Time Happens* (New York: Palgrave MacMillan, 2015), pp. 91–100. https://doi.org/10.1057/9781137410276_7

McCulloch, Warren S., 'A Historical Introduction to the Postulational Foundations of Experimental Epistemology', in F.S.C. Northrop and H.H. Livingston (eds), *Cross-Cultural Understanding: Epistemology in Anthropology* (New York: Harper & Row, 1964), pp. 39–50

McLuhan, Marshall, *Understanding Media: The Extensions of Man* (New York: McGraw-Hill, 1964)

McLuhan, Marshall and Eric McLuhan, *Laws of Media: The New Science* (Toronto: University of Toronto Press, 1988)

McLuhan, Marshall and Bruce R. Powers, *The Global Village: Transformations in World Life and Media in the 21st Century* (Oxford: Oxford University Press, 1989)

Meyer-Eppler, Werner, 'Elektronische Kompositionstechnik', *Melos*, 20, no. 1 (1953), pp. 5–9

Mira, Christian, 'I. Gumowski and a Toulouse Research Group in the "Prehistoric" Times of Chaotic Dynamics', in Ralph Abraham and Yoshisuke Ueda, *The Chaos Avant-Garde: Memories of the Early Days of Chaos Theory* (Singapore: World Scientific, 2000), pp. 95–198. https://doi.org/10.1142/9789812386472_0008

Miranda, Eduardo Reck and James L. Weaver, 'Introduction to Quantum Computing for Musicians', in Eduardo Reck Miranda (ed.), *Quantum Computer Music: Foundations, Methods and Advanced Concepts* (Cham: Springer, 2022), pp. 1–20. https://doi.org/10.1007/978-3-031-13909-3

Miyazaki, Shintaro, 'Algorhythmics: Understanding Micro-Temporality in Computational Cultures', *Computational Culture*, 2 (2012), http://computationalculture.net/algorhythmics-understanding-micro-temporality-in-computational-cultures [accessed 26 June 2025]

Moles, Abraham A., *Théorie de l'information et perception esthétique* (Paris: Denoël, 1958)

Morton, Timothy, *Hyperobjects: Philosophy and Ecology after the End of the World* (Minneapolis and London: University of Minnesota Press, 2013)

Nake, Frieder, 'Das doppelte Bild', *Bildwelten des Wissens: Kunsthistorisches Jahrbuch für Bildkritik*, 3, no. 3 (2006), thematic issue 'Digitale Form', ed. by Margarete Pratschke, pp. 40–50. https://doi.org/10.1515/9783110547375-005

Noack, Pit and Sophia Sanner, *Künstliche Intelligenz verstehen: Eine spielerische Einführung* (Bonn: Rheinwerk, 2nd ext. edn 2023)

Obvious, *Artificial Intelligence for Art*, https://obvious-art.com/wp-content/uploads/2020/04/MANIFESTO-V2.pdf [accessed 29 July 2024]

Ochsner, Beate and Robert Stock, 'Das Hören des Cochlea-Implantats', *Historische Anthropologie*, 22, no. 3 (2014), pp. 408–25. https://doi.org/10.7788/ha-2014-0308

Oever, Annie van den and Andreas Fickers, 'Doing Experimental Media Archaeology: Epistemological and Methodological Reflections on Experiments with Historical Objects of Media Technologies', in Ben Roberts and Mark Goodall (eds), *New Media Archaeologies* (Amsterdam: Amsterdam University Press, 2019), pp. 45–68. https://doi.org/10.2307/j.ctvcj303s.6

Panayotov, Stanimir, '"To Engineer the Time by Other Means": Interview with Luciana Parisi' [2016], *Academia*, https://www.academia.edu/30636968/_To_Engineer_the_Time_by_Other_Means_Interview_with_Luciana_Parisi [accessed 16 July 2025]

Parikka, Jussi, *A Geology of Media* (Minneapolis and London: University of Minnesota Press, 2015). https://doi.org/10.5749/minnesota/9780816695515.001.0001

Parisi, Luciana, 'Automated Architecture: Speculative Reason in the Age of the Algorithm', in Armen Avanessian and Robin Mackay (eds), *#accelerate#: the accelerationist reader* (London and Berlin: Urbanomic and Merve, 2014), pp. 409–24

Pask, Gordon, 'A Comment, a Case History and a Plan', in Jasia Reichardt (ed.), *Cybernetics, Art, and Ideas* (Greenwich, CT: New York Graphics Society, 1971), pp. 76–99

Pearson, S.O. and H. St G. Anson, 'Demonstration of Some Electrical Properties of Neon-filled Lamps', *Proceedings of the Physical Society of London*, 34 (1921), pp. 175–76. https://doi.org/10.1088/1478-7814/34/1/435

Peirce, Charles S., 'Lectures on Pragmatism', in idem, *The Collected Papers of Charles Saunders Peirce*, vol. V: *Pragmatism and Pragmaticism* [1934] (Cambridge: The Belknap Press of Harvard University Press, 2nd edn 1960), pp. 14–212

Peirce, Charles S., 'Pragmatism and Abduction', in idem, *Collected Papers of Charles S. Peirce*, ed. by Charles Hartshorne and Paul Weiss (Cambridge, MA: Harvard University Press, 1974), pp. 112–31

Peirce, Charles S., 'Logical Machines', in idem, *The New Elements of Mathematics*, vol. III/1: *Mathematical Miscellanea* (The Hague / Paris and Atlantic Highlands, NJ: Mouton / Humanities Press, 1976), pp. 625–32

Peters, Johannes, 'Entropie und Information', *Physik in unserer Zeit*, 1, no. 5 (1970), pp. 162–67. https://doi.org/10.1002/piuz.19700010505

Pias, Claus (ed.), *Cybernetics / Kybernetik: The Macy Conferences 1946–1953*, vol. 1 (Berlin: diaphanes, 2003)

Pink, Sarah, Elisenda Ardèvol and Débora Lanzeni (eds), *Digital Materialities: Design and Anthropology* (London: Bloomsbury Academic, 2016)

QRT [Leiner, Markus Konradin], 'Tekknologic als Tekknowledge', in idem, *Tekknologic, Tekknowledge, Tekgnosis: Ein Theoriemix* (Berlin: Merve, 1999), pp. 7–35

Radford, Alec et al., 'Learning Transferable Visual Models From Natural Language Supervision', *arXiv.org* (2021). https://arxiv.org/pdf/2103.00020.pdf

Rancière, Jacques, *The Names of History: On the Poetics of Knowledge*, transl. by Hassan Melehy (Minneapolis: University of Minneapolis Press, 1994)

Rekveld, Joost, *Dialogues with Machines* [PhD thesis] (University of Gent, Department of Art, Music and Theatre Sciences, October 2024). https://biblio.ugent.be/publication/01JAJH6XA7BD5QC1DVKYNHV9RD [accessed 21 October 2024]

Reuleaux, Franz, 'Wie beschreibt und erläutert man am besten eine Maschine oder eine andere technische Einrichtung?', *Verhandlungen des Vereins zur Beförderung des Gewerbefleißes in Preußen*, 51 (1872), pp. 108–28

Reuleaux, Franz, *Theoretische Kinematik: Grundzüge einer Theorie des Maschinenwesens* (Braunschweig: Vieweg, 1875)

Rheinberger, Hans-Jörg, *Towards a History of Epistemic Things: Synthesizing Proteins in the Test Tube* (Stanford, CA: Stanford University Press, 1997)

Rieger, Matthias, *Helmholtz Musicus: Die Objektivierung der Musik im 19. Jahrhundert durch Helmholtz' Lehre von den Tonempfindungen* (Darmstadt: Wissenschaftliche Buchgesellschaft, 2006)

Rietbrock, Bernhard, *Alvin Lucier's Reflexive Experimental Aesthetics* (Hofheim: Wolke, 2022)

Riis, Morten, 'Where Are the Ears of the Machine? Towards a Sounding Micro-Temporal Object-Oriented Ontology', *Journal of Sonic Studies*, 10 (2015). https://www.researchcatalogue.net/view/219290/219291

Rohrhuber, Julian, 'Algorithmic Music and the Philosophy of Time', in Alex McLean and Roger T. Dean (eds), *The Oxford Handbook of Algorithmic Music* (Oxford and New York: Oxford University Press, 2018), pp. 17–40. https://doi.org/10.1093/oxfordhb/9780190226992.013.1

Rombach, Robin, Andreas Blattmann, Dominik Lorenz, Patrick Esser and Björn Ommer, 'High-Resolution Image Synthesis with Latent Diffusion Models', *arXiv.org* [2021, revised April 2022]. https://doi.org/10.48550/arXiv.2112.10752

Salvaggio, Eryk, 'Toy Cars and Cybernetic Communism', *Cybernetic Forests* (entry 28 March 2021). https://cyberneticforests.substack.com/p/toy-cars-and-cybernetic-communism [accessed 20 September 2024]

Schäffner, Wolfgang, 'Topologie der Medien: Descartes, Peirce, Shannon', in Stefan Andriopoulos, Gabriele Schabacher and Eckhard Schumacher (eds), *Die Adresse des Mediums* (Cologne: DuMont, 2001), pp. 82–93

Schulze, Malte, 'Löten', in Stefan Höltgen (ed.), *Medientechnisches Wissen*, vol. 4: *Elektronik, Elektronikpraxis, Computerbau* (Berlin and Boston: De Gruyter, 2023), pp. 176–99. https://doi.org/10.1515/9783110581805-011

Searle, John R., 'Minds, Brains, and Programs', *The Behavioral and Brain Sciences*, 3 (1980), pp. 417–57. https://doi.org/10.1017/S0140525X00005756

Serres, Michel, 'Der Mensch ohne Fähigkeiten: Die neuen Technologien und die Ökonomie des Vergessens', *Transit*, 22 (Winter 2001/2), pp. 193–206

Shannon, Claude E., 'A Symbolic Analysis of Relay and Switching Circuits', *Transactions of the American Institute of Electrical Engineers*, 57 (1938), pp. 713–23. https://doi.org/10.1109/T-AIEE.1938.5057767

Shannon, Claude E., 'A Mathematical Theory of Communication', *The Bell System Technical Journal*, 27 (July–October 1948), pp. 379–423, 623–56, reprinted with corrections. https://doi.org/10.1002/j.1538-7305.1948.tb01338.x

Shannon, Claude E., 'Presentation of a Maze-Solving Machine', in Heinz von Foerster (ed.), *Cybernetics: Circular Causal and Feedback Mechanisms in Biological and Social Systems: Transactions of the Eighth Conference, March 15–16, 1951* (New Yor: Josiah Macy Jr Foundation, 1952), pp. 173–80

Shannon, Claude, 'The Mathematical Theory of Communication', in idem and Warren Weaver, *The Mathematical Theory of Communication* (Urbana, IL: University of Illinois Press, 1963), pp. 29–125

Shannon, Claude E., John R. Pierce and B.M. Oliver, 'The Philosophy of PCM' [*1948], in N. Sloane and A. Wyner (eds), *Claude Elwood Shannon: Collected Papers* (Piscataway, NJ: IEEE, 1993), pp. 151–59

Sharpless, T. Kite, 'Mercury Delay Lines as a Memory Unit', in *Proceedings of a Symposium on Large-Scale Calculating Machinery* (Cambridge, MA: Harvard University Press, 1948), pp. 103–09

Siegert, Bernhard, 'Das Leben zählt nicht. Natur- und Geisteswissenschaften bei Dilthey aus mediengschichtlicher Sicht', in Claus Pias (ed.), *Medien: Dreizehn Vorträge zur Medienkultur* (Weimar: VDG, 1999), pp. 161–82

Siegert, Bernhard, *Cultural Techniques: Grids, Filters, Doors, and Other Articulations of the Real*, transl. by G. Winthrop-Young (New York: Fordham University Press, 2014). https://doi.org/10.5422/fordham/9780823263752.001.0001

Simondon, Gilbert, *On the Mode of Existence of Technical Objects*, transl. from the French original [1958] by Cecile Malaspina and John Rogove (Minneapolis: Univocal Press, 2017)

Soentgen, Jens, 'Atome Sehen, Atome Hören', in Alfred Nordmann, Joachim Schummer and Astrid Schwarz (eds), *Nanotechnologien im Kontext: Philosophische, ethische und gesellschaftliche Perspektiven* (Berlin: Akademische Verlangsgesellschaft, 2006), pp. 97–113

Sohl-Dickstein, Jascha, Eric A. Weiss, Niru Maheswaranathan and Surya Ganguli, 'Deep Unsupervised Learning Using Nonequilibrium Thermodynamics', *arXiv.org* (2015). https://arxiv.org/abs/1503.03585 [accessed 27 November 2024]

Sparkes, Matthew, 'First Working Graphene Semiconductor Could Lead to Faster Computers', *New Scientist*, 4 January 2024. https://doi.org/10.1016/S0262-4079(24)00053-8

Spitzer, Manfred, *Musik im Kopf: Hören, Musizieren, Verstehen und Erleben im neuronalen Netzwerk* (Stuttgart and New York: Schatteuer, 2002)

Striano, Francesco, 'Log.icona. La co-originarietà di lógos e eikṓn riemerge nel digitale', in N. Russo and J. Mutchinick (eds), *Immagine e memoria nell'era digitale* (Udine: Mimesis, 2020), pp. 77–95

Tkaczyk, Viktoria and Christina Dörfling, 'Wireless Wiring: The BBC Empire Service, Manipulation of Time and Space, and Material Resistances', *Sound Studies*, special issue on tape recording, ed. by Peter McMurray and Andrea Bohlman (forthcoming). German version: 'Ton auf Band: Raum-Zeit-Manipulationen und Materialwiderstände im BBC Empire Service', in Christoph Borbach, Timo Kaerlein, Robert Stock, Sabine Wirth (eds), Akustische Interfaces (Wiesbaden: Springer, 2025), pp. 245–267

Tousseul, Sylvain, 'L'illusion de la pensée-machine', in François Laruelle (ed.), *Homo ex machina* (Paris: L'Harmattan, 2005), pp. 173–216

Trillo, Roberto Alonso and Marek Poliks, '*Debris*: Machine Learning, Archive Archaeology, Digital Audio Waste', *Organised Sound*, 28, no. 3 (December 2023), pp. 392–406. https://doi.org/10.1017/S1355771823000249

Turing, Alan M., 'On Computable Numbers with an Application to the Entscheidungsproblem', *Proceedings of the London Mathematical Society*, 42 (1937), pp. 230–65. https://doi.org/10.1112/plms/s2-42.1.230

Turing, Alan M., 'The Chemical Basis of Morphogenesis', *Philosophical Transactions of the Royal Society of London*, 237, no. 641 (1952), pp. 37–72. https://doi.org/10.1098/rstb.1952.0012

Ueda, Yoshisuke, 'Strange Attractors and the Origin of Chaos', in Ralph Abraham and idem, *The Chaos Avant-Garde: Memories of the Early Days of Chaos Theory* (Singapore: World Scientific, 2000), pp. 23–56. https://doi.org/10.1142/9789812386472_0003

Van Gulik, R.H., 'The Lore of the Chinese Lute: An Essay in Ch'in Ideology', *Monumenta Nipponica*, 3, no. 1 (1940), pp. 127–76. https://doi.org/10.2307/2382409

Viola, Bill, 'The Sound of One Line Scanning', in Dan Lander and Micah Lexier (eds), *Sound by Artists* (Toronto: Banff, 1990), pp. 39–54

Volmar, Axel and Kyle Stine (eds), *Media Infrastructures and the Politics of Digital Time: Essays on Hardwired Temporalities* (Amsterdam: Amsterdam University Press, 2021). https://doi.org/10.2307/j.ctv1xcxr3n

Wershler, Darren, Lori Emerson and Jussi Parikka, *The Lab Book: Situated Practices in Media Studies* (Minneapolis and London: University of Minnesota Press, 2021). https://doi.org/10.5749/9781452958408

Westphal, Henry, 'Vom Messen und Zählen: Wie "Analog" durch Mitkopplung zu "Digital" wird', in Moritz Hiller and Stefan Höltgen (eds), *Archäographien: Aspekte einer Radikalen Medienarchäologie* (Berlin: Schwabe, 2019), pp. 101–14. https://doi.org/10.31267/978-3-7574-0029-3

Whitehead, Alfred North, *Science and the Modern World* (New York: Free Press, 1967)

Wiener, Norbert, 'Computing Machines and the Nervous System', in idem, *Cybernetics: Or Control and Communication in the Animal and the Machine* [*1948] (Cambridge, MA: MIT Press, 2nd edn 1962), pp. 116–32. https://doi.org/10.1037/13140-005

Wittje, Roland, 'The Electrical Imagination: Sound Analogies, Equivalent Circuits, and the Rise of Electroacoustics, 1863–1939', *Osiris*, 28, no. 1 (2013), pp. 40–63. https://doi.org/10.1086/671362

Index

www.ingramcontent.com/pod-product-compliance
Ingram Content Group UK Ltd.
Pitfield, Milton Keynes, MK11 3LW, UK
UKHW041644031026
15587UKWH00001B/480

9 781804 132326